THE WEB OF LIFE

Part 1
Web of Life

by John Oates

ISBN: 84-382-0017-6. Dep. Legal: 26.891-1975

Printed and bound in Spain by Novograph
S.A., and **Roner** S.A., Crta de Irun, Km.12,450,
Madrid–34.

Series Coordinator	Geoffrey Rogers
Art Director	Frank Fry
Design Consultant	Guenther Radtke
Editorial Consultant	Donald Berwick
Series Consultant	Malcolm Ross-Macdonald
Editors	Bridget Gibbs
	Damian Grint
	Maureen Cartwright
Research	Carol Potter
	Ann Fisher
	Enid Moore
Art Assistants	Vivienne Field
	Simon Hampshire

Contents: Part 1

Editorial Advisers

DAVID ATTENBOROUGH Naturalist and Broadcaster.

MICHAEL BOORER, B.SC. Author, Lecturer, and Broadcaster.

MATTHEW BRENNAN, ED.D. Director, Brentree Environmental Center, Professor of Conservation Education, Pennsylvania State University.

PHYLLIS BUSCH, ED.D. Author, Science Teacher, and Consultant in Environmental Education.

MICHAEL HASSELL, B.A., M.A.(OXON), D.PHIL. Lecturer in Ecology, Imperial College, London.

STUART MCNEILL, B.SC., PH.D. Lecturer in Ecology, Imperial College, London.

JAMES OLIVER, PH.D. Director of the New York Aquarium, former Director of the American Museum of Natural History, former Director of the New York Zoological Park, formerly Professor of Zoology, University of Florida.

Foreword by David Attenborough

Imagine the thrill that an explorer feels as he steps ashore on an unknown continent. Strange flowers bloom on the bushes, extraordinary birds call from the trees and creatures unlike anything he has ever seen scamper off into the undergrowth. Columbus must have marveled at such sights; so must Captain Cook—and so did one 17th-century Dutch draper! But the draper did not travel in a ship manned by a large crew, nor did he journey for thousands of miles. He looked into his new world without leaving his house in Delft and with the aid of only a small piece of glass. His name was Anton van Leeuwenhoek.

The world he discovered was full of wonders. He peered into a drop of pond water through the single-lensed microscope he had built and watched a tiny filament seize small objects with its tentacles. The filament was a minute creature related to a sea anemone.

In the 300 years since then, other workers with increasingly powerful and sophisticated microscopes have explored further and further. Dr. Toomer and Dr. Cane describe some of the discoveries—great pastures of minute plants floating in the sea; tiny hunters, propelled by waving threads, that pursue their minuscule prey; organisms that live by consuming the dead or by invading the bodies of the living. There are some that cause mosaics on plant leaves, some that produce fevers in human beings, some that turn milk into cheese. The variety of these normally invisible beings seems unlimited, yet they pervade every corner of the world.

In Leeuwenhoek's time, the basic processes of life were only dimly understood. It was thought quite possible that dust could coagulate and form a flea and that a wheat grain could decompose and become a weevil. Leeuwenhoek looked at such creatures with his tiny lens and marvelously observant eye and showed that neither had had such origins and that both had developed from eggs and, in their turn, had laid them, just like other insects.

Investigations of such links between organisms and their environment are a fundamental part of the relatively new science of ecology. It might seem unbelievable that there could be any link between, say, the filling-in of a pond and the number of beetles in a cave several miles away. It becomes understandable if you know that the pond provided rich insect food for a small colony of bats living in the cave, that their feces nourished a mold growing on the cave floor, and that this in turn was the only food of the cave beetles. There is an infinity of much more complex threads in the web of life that Dr. Oates examines. His subject is a vital one. Today, because of our immense technical powers and our sheer numbers, we are interfering with the processes of nature in a more radical way than ever before. We are still a very long way indeed from understanding the full effects of what we are doing when we spray a field with insecticide, breed a new kind of domestic animal, or reduce the numbers of another animal to the edge of extinction. But if we take to heart the basic principles that ecologists are now discovering, we may still avoid the thoughtless acts that could rupture the web of life and devastate our world.

David Attenborough.

The Living Planet

Gazing into the night sky, man sees a myriad of stars beaming across the void of space. Interrupted by the gases of our atmosphere, the beams of starlight twinkle in the dark. Some of the brightest objects viewed from the darkened face of the earth are other planets circling the same star—our sun—and reflecting its brilliant light. Even brighter than these is earth's own planet, the moon; when not too greatly shadowed by the earth, the moon lights up our nights with reflected sunshine.

The moon is a particularly good reflector of sunlight, for its pale surface is barren. The only living things that can gaze at earth from the moon are creatures that have traveled there from elsewhere. But starlight does not twinkle in the lunar sky, for the moon lacks the blue, watery atmosphere that earth retains by the greater gravitational pull that its size exerts. It is earth's atmospheric coating that has allowed the development of a vast array of living things, all ultimately dependent on energy from the sun to maintain and reproduce their complex structures. One particular earth organism, calling itself "man," has now, some 4500 million years after the planet was formed, reached a complexity that enables it to construct images explaining the nature of itself and its surroundings.

By night, man may look out at the universe and attempt to explain its nature, but by day, when the brilliance of the sun obscures the stars, it is man's earthly surroundings that occupy his inquiring mind. In the cities, where so many people live today, most of the obvious features of the surroundings have been created largely by man himself. But for nearly all of the hundreds of thousands of years of man's existence there have been no cities, and other life forms and natural features have dominated the scene. These in turn may have been modified by man, but they

A lunar view of half earth. From the barren surface of the moon our planet appears as a blue-and-white disk hanging against the dark backcloth of space. Earth's characteristic coloration comes from the sun's rays being reflected from clouds and scattered by the oceans and atmosphere. It is earth's watery surface that has made possible the development of life on our planet.

are not totally under his control. In fact, man's existence depends on the many billions of organisms that share the earth's surface with him.

City man still depends on organisms living outside cities; the oxygen he breathes is produced by plants, the food he eats is derived from the bodies of plants and animals, and vegetation influences the drainage of rainfall from the land and so affects the water he drinks.

Man, other animals, and plants do not grow in isolation. The same sun energy flows through them all, and the same atoms of matter circulate constantly between them and the physical environment. Man, together with other living things and with rocks, air, and water, is linked in a complex web. How has life come into existence on this small planet, and what is the nature of the web that links its multitudes of forms?

Planet earth, in common with the rest of our solar system, probably formed as a result of the condensation of an enormous disk of dust and gas spinning through space. The sun condensed first, at the center of this disk, followed by the planets. The outer coating of earth must have become wet early in its history, perhaps as a result of water escaping from the interior to the rocky surface. As the planet, with its coating of solids, liquids, and gases, circled the sun it was subject to intense ultraviolet radiation (cut off today by ozone gas in the upper atmosphere), which, together with lightning from electrical storms, could have led to the formation of amino acids and sugars in the primeval oceans. Concentration of these molecules by condensation may have followed, producing larger, more complex molecules that were able to reproduce accurately their own ordered structure. So, sometime during the first 1000 million years of earth's history, might life have begun. Reproduction soon resulted in a huge population of similar molecules, competing with one another for the available matter and energy with which to build more of themselves.

But these complex living molecules did not remain identical. Exposed to radiation and chemicals, some underwent small changes. Early in its history, life started to vary. Not all the

The amino acids in the primitive oceans may have been joined to sugars at the sites of volcanic eruptions (below) to produce complex self-reproducing molecules. Then, according to theory, the evolution of living things could begin.

Today, much of the once-barren land surface of our planet is covered with a blanket of life. One of the thickest parts of this blanket is to be found on the temperate Pacific coast of North America, where heavy rainfall supports a dense rain forest containing some of the world's largest trees.

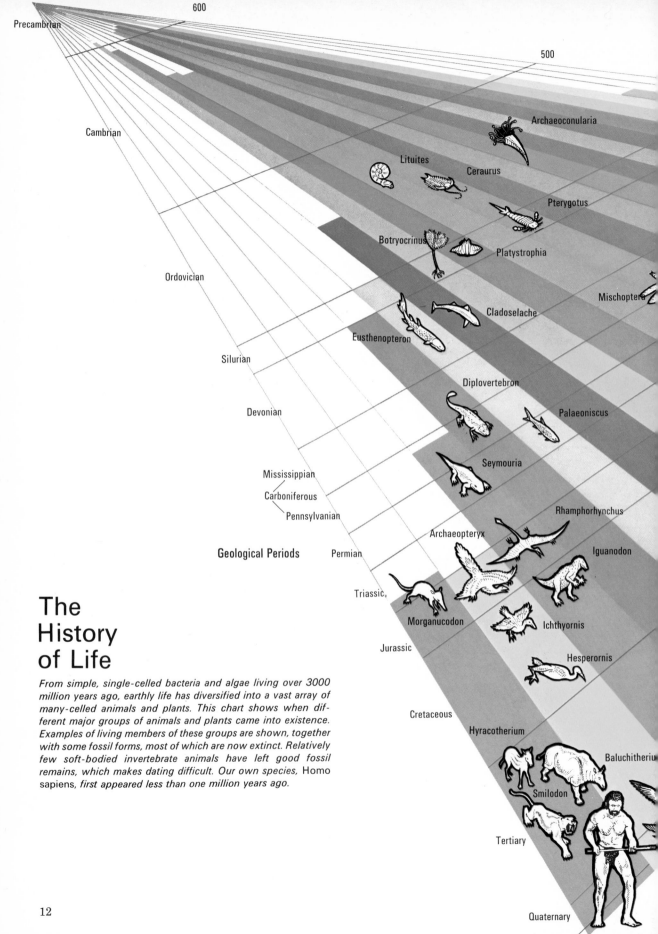

The History of Life

From simple, single-celled bacteria and algae living over 3000 million years ago, earthly life has diversified into a vast array of many-celled animals and plants. This chart shows when different major groups of animals and plants came into existence. Examples of living members of these groups are shown, together with some fossil forms, most of which are now extinct. Relatively few soft-bodied invertebrate animals have left good fossil remains, which makes dating difficult. Our own species, *Homo sapiens*, *first appeared less than one million years ago.*

Precambrian

600

500

Cambrian

Ordovician

Silurian

Devonian

Mississippian

Carboniferous

Pennsylvanian

Geological Periods

Permian

Triassic

Jurassic

Cretaceous

Tertiary

Quaternary

Archaeoconularia

Lituites

Ceraurus

Pterygotus

Botryocrinus

Platystrophia

Mischoptera

Cladoselache

Eusthenopteron

Diplovertebron

Palaeoniscus

Seymouria

Rhamphorhynchus

Archaeopteryx

Iguanodon

Morganucodon

Ichthyornis

Hesperornis

Hyracotherium

Baluchitherium

Smilodon

Modern man (*Homo sapiens*) appeared here

Mammals

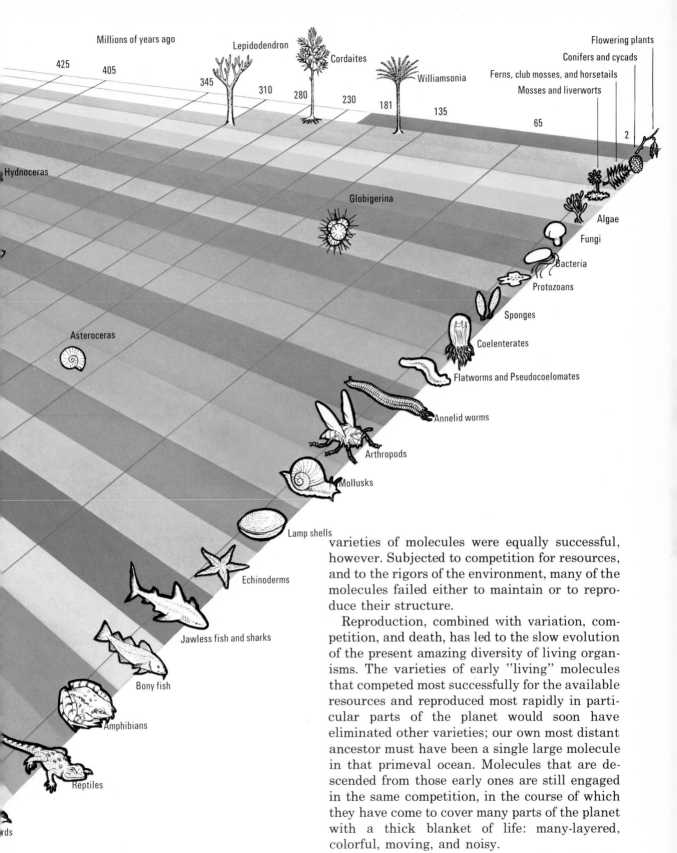

Millions of years ago

425 405 345 310 280 230 181 135 65 2

Hydnoceras

Lepidodendron

Cordaites

Williamsonia

Flowering plants
Conifers and cycads
Ferns, club mosses, and horsetails
Mosses and liverworts

Globigerina

Algae

Fungi

Bacteria

Protozoans

Sponges

Coelenterates

Asteroceras

Flatworms and Pseudocoelomates

Annelid worms

Arthropods

Mollusks

Lamp shells

Echinoderms

Jawless fish and sharks

Bony fish

Amphibians

Reptiles

varieties of molecules were equally successful, however. Subjected to competition for resources, and to the rigors of the environment, many of the molecules failed either to maintain or to reproduce their structure.

Reproduction, combined with variation, competition, and death, has led to the slow evolution of the present amazing diversity of living organisms. The varieties of early "living" molecules that competed most successfully for the available resources and reproduced most rapidly in particular parts of the planet would soon have eliminated other varieties; our own most distant ancestor must have been a single large molecule in that primeval ocean. Molecules that are descended from those early ones are still engaged in the same competition, in the course of which they have come to cover many parts of the planet with a thick blanket of life: many-layered, colorful, moving, and noisy.

Self-reproducing molecules are found in the

13

cell nuclei of all living organisms. They are called nucleic acids, and the most important of them is deoxyribonucleic acid—DNA for short.

Nucleic acids engage in the reproductive competition of life by organizing other matter around them in the form of cells, using energy derived ultimately from the sun. The *cell* was given its name in 1665 by the English scientist Robert Hooke, who described the similarity of the walls of dead cells in a piece of cork to those of a bee's honeycomb cell. The cell is a functional unit containing the machinery for maintenance, growth, and reproduction. In present-day environments DNA is unable to reproduce itself, or *replicate*, without this machinery. The cell is, therefore, the fundamental unit of life.

Most organisms consist of cells or organized groups of cells. But there is a notable exception — viruses. These are nucleic acids surrounded by nothing more than a protein coat; they can reproduce only by entering a living cell and taking over its machinery. Often, when they take over cells in our own bodies, they cause disturbances that we regard as disease. Viruses may have evolved from truly cellular organisms; they are the exceptions that prove the rule by demonstrating that nucleic acids are the hereditary and organizing molecules fundamental to life, but that the cell machinery is essential for the reproduction of the nucleic acid.

Organisms exist in related sets, or species. A species is a set of organisms that contain the same varieties of DNA molecules and are capable of exchanging parts of this hereditary, or genetic, material. By exchanging parts, variation in a population of DNA molecules of a particular variety is increased. This exchange, the basis of sexual reproduction, increases the possibility that a variety more successful in competition with other molecules will appear. For example, take one species: the ostrich. A male ostrich may mate successfully with any female ostrich, but not with an emu, an eagle, or a sparrow. The mating process results in the production of offspring each containing a different combination of the two parents' DNA. Some of the ostriches will be better fitted for the life ahead of them than others. The most successful ostriches tend to produce the best-adapted offspring.

Life may have arisen more than once on the primeval earth, but only one form of life, based on the protein DNA, has been successful. This form has evolved by the process of natural selection into an enormous number of very different species. A parallel may be found in the development of human machines.

During the 1800s many attempts were made to produce a self-propelled road vehicle, but it was not until 1891 that the first really successful automobile was made.

Early motor vehicles were relatively simple, inefficient, and rare. But competition between manufacturers and efforts to adapt cars to the public's needs has made them increasingly more complex, and more efficient both in the utilization of energy and in performance. A whole range of specialized vehicles to perform different tasks has "evolved": cars for carrying only a few people, buses for many passengers, trucks for carrying goods, bulldozers for shifting earth, tractors for pulling machinery, and so on. The number of automobiles has increased vastly since the early days, and they have spread throughout the world. Fortunately, they cannot yet reproduce themselves.

Although living organisms are varied and abundant on earth, they are still linked together; they probably share a common ancestry, they certainly share a common habitat—the thin biosphere covering the earth's surface—and they compete for the same sources of energy and materials. They are not distinct, totally independent entities, but are especially organized parts of the earth's coat that interact constantly with one another, as well as with the nonliving parts of the coat. They form a web of life that has become increasingly complex with the passage of time, and that links every living thing on earth.

Ecology is the branch of science that looks at the interactions of living organisms in this web with one another, and with their surroundings. The word "ecology" comes from the Greek words *oikos*—a house, and *logos*—the way a subject is spoken about or treated. Ecology is the subject that speaks about the "houses" of living things — the study of how organisms interact with the various living and nonliving parts of the environment in which they live.

A fundamental element in the web of life is the food chain. When a hunting animal kills its prey and eats, it is acting as one of the links in such a chain. We can illustrate the principle of a food chain by considering the Atlantic herring, which is an important part of man's diet in many countries. The herring's main food in the North Sea is the small shrimplike animal called *Calanus*

The machinery of life. Right: a model of DNA, the complex self-reproducing molecule that occurs in the cell nucleus and contains coded information that regulates the development and functioning of the cell. In the developing embryo of a many-celled organism, such as the human fetus (below, center) cells become different (forming, for instance, blood, bone, nerve, and cartilage cells, below left and right) due to the selective functioning of their DNA. DNA is contained in the threadlike bodies called chromosomes *(far right).*

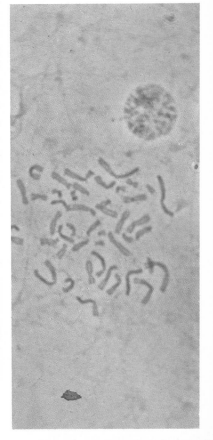

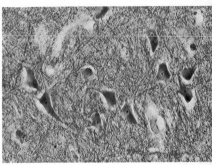

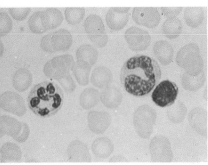

Blood Cells

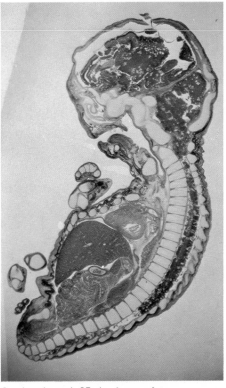

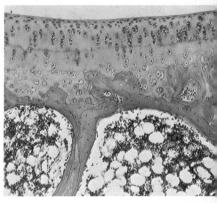

Nerve cells

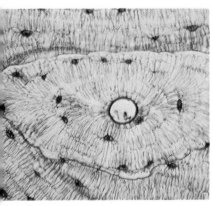

Bone

Section through 65-day human fetus

Cartilage

finmarchicus. Herrings congregate to feed in areas where *Calanus*, which lives in the upper layers of the sea, occurs in large numbers. *Calanus* itself feeds on tiny floating plants, called *phytoplankton*, which it sieves from the water using especially adapted mouthparts. These microscopic plants are probably similar to some of the earliest cells of the earth's primeval oceans. Although minute, they are so abundant in some waters that they tinge the sea green.

Phytoplanktonic plants, in common with most land plants, are green because their cells contain the pigment *chlorophyll*, which traps the energy of sunlight. Plants use this energy to build their own complex structure from simple materials absorbed from their surroundings. Water and carbon dioxide are the raw products used in this process, which we call *photosynthesis*. But plants also require certain mineral salts, and these are only abundant in surface sea water in relatively few parts of the world, of which the North Sea is one. The surface waters of the Mediterranean lack large quantities of dissolved minerals, and their clear blue appearance indicates the low density of phytoplankton that is the result.

Through food chains, the sun's energy and simple materials from the earth's surface are ultimately incorporated into our own bodies.

These food chains may be very complex. Man, for instance, is involved in a great many food chains, on land and sea; nor is he strictly at the "end" of a chain. For, when organisms in a food chain excrete or when they die and their bodies are decomposed by bacteria and fungi, material is returned to the environment, ready to re-enter another food chain.

Energy, however, flows through the chain and is not recycled. Trapped in the form of sunlight by green plants at the start of the food chain, it is given off as heat at points all along the chain and is ultimately lost from the planet by radiation into space. The constant flow of energy to the earth's surface from the sun replaces this heat.

Throughout food chains, animals and plants are involved in competition for limited resources, just as man competes for herrings with other fish-eating animals, such as seabirds. This competition results in adaptation through the process of evolution. Competition may also lead to cooperation. Members of the same species frequently cooperate in such a way that they are individually more successful in obtaining food and producing offspring. But there are also many examples of cooperation between members of *different* species.

Food chains, the circulation of matter, the flow of energy, competition, cooperation, adaptation—these are the basic components in the living web that distinguishes the busy surface of the earth from the empty landscape of the moon. They are elements that feature in the lives of all plants and animals, man included.

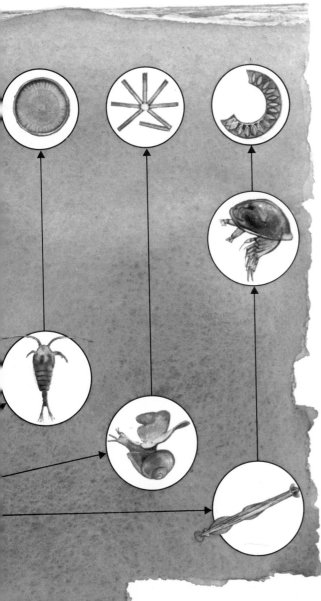

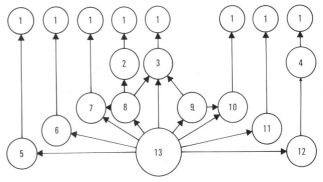

Food Web of the Atlantic Herring

1 Diatoms and dinoflagellates		Phytoplankton *(floating plants)*
2 Branchiopod crustacean	*(Evadne)*	
3 Copepod crustacean	*(Calanus)*	
4 Barnacle larva	*(Balanus)*	
5 Tunicate	*(Oikopleura)*	
6 Copepod crustacean	*(Centropages)*	
7 Euphausid crustacean	*(Nyctiphanes)*	Zooplankton *(floating animals)*
8 Amphipod crustacean	*(Hyperia)*	
9 Sand eel	*(Ammodytes)*	
10 Copepod crustacean	*(Temora)*	
11 Sea butterfly	*(Spiratella)*	
12 Arrow worm	*(Sagitta)*	
13 Adult herring	*(Clupea)*	

Capturing herring in a purse seine net. Herring, eaten by several marine predators and man, are just one link in a complex web of feeding relationships. Most of the minute floating animals (zooplankton) on which herring feed eat several different kinds of microscopic floating plants (phytoplankton) but mainly those called diatoms; *some, however, eat other tiny floating animals.*

17

Food Chains

With his insatiable need for fuel to drive an industrial society, man has found ways of generating electricity using directly the light energy radiated by the sun, which is a gigantic nuclear reactor. But to obtain any useful amount of power, he has to cover huge areas with solar cells to trap and convert the light energy. In so doing, man imitates, very inefficiently, what plants have been doing unobtrusively for many millions of years.

Not all plants do it; mushrooms and other fungi, for instance, are an exception. But every *green* plant is a combined solar power station and factory. It has been estimated that, each year, the green plants absorb and convert an amount of sun energy equivalent to the production of 2000 million large power stations. They accomplish this amazing feat because they are able to trap light energy and then use it to join together simple molecules, gathered from their surroundings, to form glucose. This is then converted into starch, an energy store, which the plants can tap as required to provide the power for building, maintaining, and reproducing their structures — the production process based on light energy and known as photosynthesis. Because of its importance as the starting point of all food chains it is fundamental to life on earth.

The raw materials joined together by photosynthesis are carbon dioxide and water. Land plants take in carbon dioxide from the atmosphere—this gas constitutes only about 0.03 per cent of the atmosphere—and water from the surface on which they grow. Submerged water plants utilize the carbon dioxide dissolved in their fluid habitat.

Photosynthesis takes place in minute rounded structures known as *chloroplasts*, which are less than 0.00016 of an inch in length. Chloroplasts contain molecules of the green pigment chlorophyll, which absorbs light energy. In the chloroplast this light is first converted into chemical

Rain-forest vegetation along the Manu River in southeast Nigeria throws up a thick wall of green chlorophyll pigment to intercept the sun's rays. The pigment traps light, whose energy is used to convert carbon dioxide from the air, and water drawn from the soil, into glucose, which is an abundant energy store.

Above: the pads of water lilies act in the same way as the horse-chestnut leaves. Under the surface of this pond, floating mats of algae are trapping the sunlight that filters through, relying for photosynthesis on carbon dioxide dissolved in the water.

Above right: single-celled floating algae are typical plants. They exist without complex root systems and can therefore utilize light falling on the surface of deep water. Where abundant, algae virtually form a solution of chlorophyll in the water.

energy, which is then used to combine carbon dioxide and water. The presence of· abundant chlorophyll is obvious in any vegetation; forests, grasslands, and surface waters are colored green by it. The pigment appears green because most of the light it absorbs is blue and red. What the eye sees is the green and yellow light that is reflected by the pigment.

Leaves are the greenest parts of most land plants; they are the parts especially equipped for the role of photosynthesis, and most of them have flattened surfaces that present a broad area to incoming sunlight. This broad surface is necessary not just to trap sunlight, but also to gather carbon dioxide. Hundreds of tiny pores, called *stomata*, pierce the surface. Through them gases, including water vapor, diffuse between the cells and the atmosphere.

The chloroplasts of leaves are usually concentrated near the waterproof upper surface, whereas stomata are concentrated on the lower surface. The effect of this arrangement is to increase efficiency and to help reduce the loss of water by evaporation; evaporation could be extreme if large numbers of stomata opened through an upper surface exposed to bright sunlight. In some parts of the world, flat, broad leaves are a great

disadvantage because they cause too much water loss. In deserts, and in areas where water is frozen in the soil for much of the year (making it unavailable to roots), many plants have leaves rounded in cross section, with a thick waterproof covering on all surfaces. The effect of this is to reduce water loss by evaporation.

Leaflike structures are essential for photosynthesizing land plants, which must expose their chlorophyll to the light. Photosynthesis in water presents different problems. Because water absorbs light, adequate amounts of light are available only in the upper layers of ponds, lakes and oceans. Where the water is deep, plants must float if they are to stay in the light. In a watery medium, where a plant is surrounded by its photosynthetic requirements, small size increases efficiency. This is because a small structure has a large surface area in relation to its volume. This aids the absorption of materials and helps to keep the plant afloat, so that complex root and support systems are unnecessary. Most plants in the open waters of lakes and oceans are therefore microscopic algae—*phytoplankton.*

The raw materials of photosynthesis are not all that plants need in order to build their structures. Many of the molecules that form the bodies of plants, and that are vital to their functioning, contain other elements. For instance, each molecule of chlorophyll itself contains one magnesium and four nitrogen atoms in addition to carbon, hydrogen, and oxygen, and all proteins contain nitrogen and some also contain sulfur.

The majority of photosynthetic land plants obtain these extra materials from the soil by absorbing them into their extensive root systems. Even though nitrogen gas is a major component of the earth's atmosphere, green plants cannot draw nitrogen directly from the air but must get it either from nitrate salts in the soil, or from bacteria growing in their roots. These bacteria have the ability to fix nitrogen from the air and convert it into compounds that the green plants can utilize. Floating water plants absorb minerals directly from their surroundings. For this reason the growth of phytoplankton is very often inhibited by a low concentration of necessary minerals

Browsing behavior of three different African animals: an elephant tears branches off an acacia with its trunk, green locusts snip leaves with their mandibles, and the long-necked giraffe employs strong muscular lips especially adapted for grasping.

in the upper layers of the sea and many lakes.

Ecologists call plants that synthesize their own food materials *primary producers*. The amount of food that they produce is influenced by the quantities of light, carbon dioxide, water, and essential minerals available, and by temperature conditions. Vegetation growing on rich alluvial plains formed from river silt, and the algae living in coral reefs, are two of the earth's most productive communities, producing over half an ounce of material daily in every square yard of growing space. However, very large areas of the earth's surface are covered by

Primary consumers that feed on living plant material are either grazers, such as these grass-eating topi antelope on an East African plain (left), or browsers, such as the leaf-eating animals above. Each species has adapted to its particular food source.

open ocean and desert that produce less than 0.005 of an ounce of material per day in every square yard.

Despite their efficiency, plants are at best able to convert into photosynthetic products only about 20 per cent of the light energy they receive. Many land plants fail to achieve even this level of production because of a lack of carbon dioxide and water, and lack of minerals, especially phosphorus, reduces the efficiency of most water plants. However, even though plants are far from being completely efficient energy converters, it has been estimated that worldwide photosynthesis produces as much as 164,000 million tons of organic material each year.

Because photosynthesis is a quiet process, invisible to man, it is not obvious that it is occurring. Often, food production is apparent only when a plant lays down stores to last it through

23

Above: leaf-cutting ants carry leaf fragments from tropical forest vegetation back to their nest. The ants do not feed directly on the leaves but on the fungus that grows on the chewed-up fragments after they have been stored in the ants' underground nest.

Below left: yellowtail moth caterpillar. In some forests the larvae of butterflies and moths eat more tree foliage than any other primary consumer. Below right: the Chrysomela *belongs to a group of beetles that specializes in eating leaves.*

an adverse period during which little photosynthesis can occur—such as the winter—or when it provides for its offspring by forming the food reserves present in many seeds. Human civilization itself perhaps began when man started to exploit just such a form of production—the seeds of grass plants, or grain.

The energy incorporated in plant material by photosynthesis has long been a vital resource for mankind in another way. Plants have kept generations of camp fires burning, and in modern times human society has become increasingly dependent on reserves of solar energy that were trapped by plants in the past—the fossil fuels of coal, gas, and oil, and the more recently formed peat. These reserves are the products of millions of years of photosynthesis, but at the present rate of exploitation they will be largely dissipated in a very few decades.

Plants, as we have seen, are the primary producers of food. The next link in the chain is the primary consumer—the *herbivore*, or plant-eater. Most primary consumers are animals, for animals cannot make their own food material out of solar energy, but must rely on material built up by plants. Whereas primary consumers feed directly on plants, consumers further along the food chain—predatory animals such as sharks, foxes, and tigers—have to obtain their plant material indirectly, by eating other consumers. Man himself is both a herbivore and a predator, as are many other omnivorous animals such as most bears and many monkeys.

The consumption of food is often more obvious than food production, but even so it is not always immediately apparent to the human observer. The small crustacean browsing on the phytoplankton of the ocean and the caterpillar nibbling at the leaves of forest trees often remained unnoticed. But on a fertile, grassy plain, where plants are concentrated in a thin layer in a form ideally suited to exploitation by large animals, consumption is clearly apparent. For instance, a herd of antelope grazing in the African savanna, bison on the American prairie, or cattle and sheep anywhere—all these provide striking illustrations of primary consumers that are feeding on plant production.

Land plants usually consist of a number of distinct parts performing special tasks: leaves for photosynthesis, roots both to gather water and minerals and to anchor the plant, stems to hold the leaves in the air, and special organs for storage and reproduction. Some herbivorous animals eat all of these parts, but most feed on a limited number of parts from a limited number of species. There are leaf-eaters, flower-eaters, fruit-eaters, seed-eaters, stem-eaters, and sap-suckers. Of these, leaf-eaters are the most numerous of the consumers.

In natural terrestrial habitats, most leaves grow either at ground level, in the form of grasses, or else on trees and bushes. These two major sources present different problems to consumers. On open grassland, it is best to be either large and able to run from danger, or small and able to hide. Large grazing mammals, such as horses and antelopes, have long legs, which enable them to stand at the ready for flight while their heads are lowered for food gathering. Small grazing mammals, such as the "prairie dogs" of North America, are efficient burrowers. And many grass-eating insects are camouflaged to look like blades of grass, or else escape from their enemies by taking to the wing.

Browsing on the leaves of bushes and trees requires slightly different tactics. Large browsing mammals generally have even longer legs and necks than grazers, enabling them to reach a food source that may be far from the ground. In parts of the African savanna, where the grassland is dotted with trees, live giraffes and giraffe-necked antelopes (or gerenuks) whose long necks enable them to reach up high for the leaves of trees and bushes.

Most browsers, however, are small animals, for the great bulk of tree leaves and buds grow out of the reach even of giraffes, whose shape in any case prevents them from moving with ease through a forest. The leaves of tall forest trees can be reached only by climbing or flying, and so the majority of leaf-eating animals are insects. Insect larvae (caterpillars and grubs) are often greater consumers of leaves and shoots of trees than are any other animals although a number of mammals may compete with them. Notable among these competitors are the sloths of Central and South America, leaf-eating monkeys of tropical Asia and Africa, and the koala of Australia. In the warmer latitudes where these animals live, trees bear edible leaves throughout the year so there is no shortage of this type of food. But in northern temperate forests the insects have little competition because mammals eating nothing but the leaves of trees would be starved of food for half the year. Insects are

The sunbirds of Africa and Asia, such as this African Mariqua sunbird, are nectar-feeders, gathering their food with long slender tongues protected by a curving bill. Unlike the hummingbirds of the Americas, sunbirds usually perch to feed.

better able to resist such a situation, by becoming dormant, than are mammals.

In order to get their food, leaf-eaters—both grazers and browsers—must have mouths or limbs adapted to plucking or cutting leaves. For instance, leaf-eating insects have jaws that snip like scissors, slugs and snails browse by licking leaves with a *radula* (a tongue that bears horny teeth), turtles grab with horny beaks, and grazing mammals often have cutting incisor teeth that shear grass like the blades of a lawnmower. Many browsing mammals—giraffes, tapirs, moose, and elephants—have elongated lips or snouts, which enable them to pluck leaves in bunches from bushes and trees and to strip nutritious bark. In contrast, leaf-eating monkeys gather leaves with grasping fingers.

Once the food has been taken into the body it must be digested. Leaf cells, like all plant cells, are surrounded by a tough supporting material called cellulose. Cellulose cannot be easily digested by most animals, for their guts do not produce the right enzymes to break up the material. To overcome this problem, leaf-eating animals usually grind their food into tiny pieces, so fragmenting many of the leaf cells and making their contents available for digestion. Mammals do this with teeth that bear complex grinding surfaces, but insects and a few crustaceans have structures inside their guts that perform the same function. The guts of a great many herbivorous animals harbor a rich population of micro-organisms—bacteria and protozoa—which manufacture cellulose-digesting enzymes. For instance, in the stomachs of ruminating mammals there is a large chamber in which the cellulose of ground-up leaves is fermented in a bacterial "soup."

Some consumers rely on the reproductive parts of plants—flowers and fruits—for food. These parts form neatly packaged, concentrated food

This bumblebee foraging for nectar on the flowering head of a thistle may bring about the fertilization of the next thistle it visits because of the pollen dusted on its body.

A hovering European hummingbird hawk-moth probes a flower for nectar. This is the counterpart in the insect world of the feeding behavior of American hummingbirds.

sources; but over much of the earth, flowers and fruits are produced during only a short season, when warmth and moisture conditions are suitable. Animals that eat only flowers and fruits are for this reason typically tropical; in the tropics flowers and fruits of one kind and another can often be found the year around. Outside the tropics, flower- and fruit-eaters must enter suspended animation or, as do some bees, rely on stored supplies during the time of year when their food is not available. Others migrate to more favorable areas, and still others must change their diet to survive, as do many temperate birds, such as thrushes.

Pollen, nectar, and fruit are difficult to gather. They do not grow in dense carpets as leaves often do. Most flower- and fruit-eaters are insects and birds that can fly rapidly between isolated food sources and reach foods that are often suspended in space at the end of a stalk. Bird beaks are good at dealing with these small food packages, because they enable the birds to reach into flowers and to pick off small fruits. But beaks are relatively poor tools for leaf-gathering, and comparatively few birds include leaves in their diet.

Many plants have developed lures that encourage animals to feed on or come close to their reproductive structures and to assist the plant. For instance, they may attract insects with the bait of nectar. In seeking out the nectar, an insect will brush against the pollen-bearing anthers of a flower, dusting pollen on to its body and then inadvertently carrying the pollen to fertilize the next flower it visits. Similarly, when an animal eats a fruit, it often consumes the seeds it carries. These seeds, resistant to digestion, pass undamaged through the animal's gut and may well be deposited later in a suitable site far from the parent plant, thereby aiding the dispersal of the plant species.

Only a few consumers are especially adapted to eating other parts of living land plants. The majority of wood-eaters feed on dead or dying wood, although a few beetle larvae eat living tree-trunks. Because digestion of wood is difficult, some insects, such as that causing the Dutch elm disease widespread in Britain, culture fungi that digest wood. The insects then feed on

the fungus. A number of animals feed on stems that are not particularly woody: the giant panda, the coypu, and the African "cutting-grass' (or cane rat) all favor the stems of large grasses such as bamboo or reeds. Many rodents, such as beavers and squirrels, as well as other mammals include a great deal of tree bark in their diet.

The sap that carries food materials around plants is a food source that is exploited by hordes of insects, such as aphids and mosquitos. These insects have mouthparts adapted to piercing through plant tissues into the sap, rather like a hypodermic needle.

Finally, there are a number of animals that consume the root systems of land plants. Pigs

Right: the North American gray squirrel. Tree-living squirrels are found in most forested parts of the world except Australia. Their diet includes fruits, shoots, and large numbers of nuts, which they gnaw through with sharp, well-developed incisor teeth.

Right: a young bullfinch feeds on the berries of a honeysuckle plant. The brightly colored berries have probably evolved to attract fruit-eating animals, who will disperse the seeds in the fruit. Bullfinches live on buds, insects, fruits, and, especially, on seeds from the woodlands of Europe and northern Asia.

Below: the Australian spectacled flying fox is a species of large bat that feeds at night on flowers and fruit. Flying foxes, or fruit-bats, occur throughout Southeast Asia and the islands of the Indian and West Pacific oceans.

"root" with their snouts through the upper layers of the soil, and mole rats attack from below, burrowing through the ground in Mediterranean regions and the drier parts of Africa. Smaller, but of no less importance, are the huge numbers of nematodes and insects, particularly larvae, that attack and consume root systems.

As a protection against ravagers, notably those attacking their aerial parts, many plants have evolved protective devices. A large number display spines that deter sensitive-lipped browsers. Some produce irritants. Others have an unpleasant taste. And there are others that are actually poisonous.

Land plants do not always satisfy all the requirements of primary consumers. To maintain themselves and to reproduce, these consumers sometimes have to get extra minerals and water from other sources. Sodium, one of the most important elements in animal bodies, is scarce in many plants, and animals must resort to salt licks in order to obtain an adequate supply. In some areas the numbers of plant-eating mammals may be limited not by the amount of food available, but by the availability of drinking sites and salt licks. This applies especially to elephants, whose requirements are prodigious. Both African and Asian elephants must have access to permanent water sources, and they commonly visit sites where they eat mineral-rich soil, often excavating large holes in the process.

Primary consumers living in water have to overcome problems very different from those facing land-living consumers. The primary producers, or phytoplankton, on which they feed are dispersed through the upper layer of the water, and there is no concentrated food source such as leaves or fruit. In the open water of lakes and oceans most primary consumers feed by filtering these microscopic plants from the fluid in which they float. Most of the consumers are small themselves, and float among the plants; they are known as the *zooplankton*. Crustaceans and, to a lesser extent, *tunicates* (animals that consist

The warthog of the African savanna commonly feeds in a kneeling position. It obtains a large part of its diet by digging for roots and tubers, shuffling along on its knees and turning up the soil of the baked savanna with powerful snout and tusks.

A male ostrich looks on while a giraffe reaches down to take soil from a "lick." Salt licks are of great importance to large herbivorous, or plant-eating animals, for most of the plants they eat contain only small quantities of essential minerals, such as sodium. The carnivores, or meat-eaters, on the other hand, get enough salt from the flesh and blood of their herbivore prey.

largely of jelly encasing a gut) predominate in the marine zooplankton, and these small animals feed by setting up currents of water that draw the plants through filters where they are trapped. Fresh waters also support a large number of planktonic crustaceans, especially *cladocerans* such as the "water fleas."

Waters close to shore, both fresh and salt, are rich in dissolved minerals, and often contain large quantities of phytoplankton. Because of this, an abundance of non-floating animals usually sits on and near the shore, sometimes buried in mud and sand, and sometimes attached to rocks. Included among these animals are shell-fish, sucking in water and filtering the phytoplankton from it, and barnacles, sweeping the water with their feathery appendages.

Not all the consumers of plant plankton are small. Lesser flamingos feed on plant plankton where it grows profusely in warm inland waters, using their tongues as pumps to suck in water and pumping it out through the filters that fringe their upturned bills. Some fish also feed primarily on phytoplankton, trapping the plants by driving water over the gills, to which filters—called *gill rakers*—are attached.

Not all water plants are planktonic. In shallow water there is often a rich growth of larger, attached plants. The animals that eat these plants have much in common with the browsers on land; though most of them are completely aquatic, some—geese for instance—actually live on land. Green turtles and Galápagos marine iguanas browse on the submerged fronds of aquatic plants, and mollusks all over the world rasp at non-planktonic vegetation. One unique group of mammals is totally adapted to a life of browsing: the large-lipped sea cows, which live in shallow, tropical waters—the manatees in estuaries on either side of the Atlantic, and the dugong on the coasts of the Indian and West Pacific oceans.

So, by this first link in the food chain, plant plankton and attached aquatic vegetation are incorporated into the bodies of the primary consumers, the herbivores. The primary consumers themselves are eaten by secondary consumers—predatory animals that eat plant production once removed. Until the coming of man the larger herbivores, such as elephants and whales, were rarely killed and eaten by predators (possibly their large size evolved as a means of protection against predators), but most other herbivores are preyed on by predators of some sort. Just as a huge variety of primary consumers feed on a similar variety of plants and plant structures, so there exists a wide range of predatory animals that specialize in catching and consuming different types of herbivores.

On land, predators are characteristically alert and active animals, with senses highly developed for detecting their prey, limbs for seeking it out, claws, teeth, or other weapons for catching it and rendering it readily eatable. Animal flesh is more

Above: tiny sea squirts—ranging from the size of a pea to about four inches high—live under water on algae-covered rocks close to the seashore, and suck in water through a "siphon," filtering off particles of plankton and other matter, and then expelling the water from a second siphon. It has been estimated that a sea squirt can filter as much as 30 pints of water in a day.

Right: with bills upturned in the water, lesser flamingos sieve tiny floating plants from the mineral-rich waters of Kenya's Lake Nakuru. White-necked comorants nest in the trees above them, but the flamingos themselves breed elsewhere in the Rift valley.

easily digested than plant material, and it represents an even more concentrated food source, rich in energy, protein, and mineral. But animals are not spread in dense carpets as plants often are, and they can move.

Plants first spread extensively over the land some 400 million years ago. This provided a food source for insects and their relatives: probably the most important terrestrial primary consumers of ancient times. Early amphibians, the ancestors of modern frogs and salamanders, soon flopped out of the water on to the land, the first of many backboned animals—the vertebrates—to exploit the rich food source represented by the insects. Even today, amphibians feed mostly on insects and other invertebrates. With their long, mobile tongues, frogs and toads can snap up the unwary insects that their keen eyes have detected.

Soon, amphibians capable of catching, killing, and eating other amphibians appeared on land. One such group evolved into reptiles, many of which were even more efficient predators, better

A North American bullfrog swallows a southern ribbon snake, reversing the more usual feeding relationship between amphibians and reptiles. The bullfrog, which gets its name from its loud, bellowing croak, may reach a length of eight inches or more.

34

able to move and feed on land and not needing to return to water at any point, as amphibians must do to lay their eggs. Some reptiles took to a plant diet, becoming primary consumers, and were preyed on by other reptiles. From reptiles evolved birds and mammals. And these became the dominant animal groups on the extinction of the dinosaurs and their cousins, 65 million years ago.

The first birds and mammals probably fed mostly on small animals such as insects, fish, amphibians, and reptiles. Insect-eating birds (such as warblers and flycatchers) and insectivorous mammals (such as shrews) are still abundant. But as birds and mammals have spread over the earth and diversified, they have taken up many other diets. Apart from herbivores and insectivores, there are large numbers of carnivorous land birds and mammals—those that live on the flesh of other vertebrates. Eagles, hawks, owls, dogs, cats, and members of the weasel and civet families and many others all include vertebrate meat in their diet. The fangs of carnivorous mammals and the talons of birds of prey have evolved until they have become deadly and efficient weapons.

Predatory vertebrates are among the most intelligent and active animals on land; they have to be, to catch their prey. The insect-eating swifts, bird-eating falcons, and gazelle-eating cheetahs are some of the fastest-traveling animals on earth. Birds of prey and cats have especially acute vision. Members of the mongoose and civet groups have an extremely well-developed sense of smell, which allows them to track down their prey. The hearing of the insect-eating bats has been developed into part of an echolocation system that enables them to find insects in the dark. Families of lions, wolves, and hunting dogs all cooperate closely among themselves in the hunt for the hoofed herbivores that constitute a large portion of their diet. Some reptiles have developed remarkable ways to catch their mammalian descendants: rattlesnakes, for example, detect rodents with infrared heat sensors on their snouts, and paralyze them with virulent toxins that not only allow the prey to be swallowed but also make it more digestible.

It soon becomes apparent that food chains on land are extremely complex. Although the insect-eaters may be feeding directly on primary consumers, some predators eat other predators, and thus are not secondary but tertiary, or higher, consumers. The owl that eats a shrew that has

Above: a male paradise flycatcher brings an insect to its off-spring. Flycatchers specialize in catching insects on the wing; this species inhabits wooded areas of Africa and southern Asia.

Below: this European Aesculapian snake raiding the nest of a shrike is acting as at least a fourth stage in a food chain, for the young shrikes will have been fed on a diet of insects caught for them by their parents. The snake's distensible jaws allow it to swallow items of food much larger than its own head.

Right: wading birds with long bills are adapted to gathering the worms and mollusks that lie buried in the sand and mud.

Closer inshore and on the seabed—where many of the primary consumers are mollusks and worms buried in mud or sand, or attached to rocks—a different community of predators occurs. Flatfish, carnivorous whelks, cowries, starfish, and sea otters eat mollusks from the seabed, while shore birds probe the mud for worms and shellfish. And in fresh waters, fish and insects are the major secondary consumers. They, in turn, are eaten by other fish, water birds, otters, crocodiles, and some snakes. (Many of these tertiary consumers also fill the role of top carnivore.)

At the upper end of food chains, the larger carnivores rarely eat every bit of the animals they kill. Scavenging animals are commonly associated with these carnivores, eating the bits left over after the carnivores have had their fill. Hyenas, jackals, and vultures will all gather the scraps from a lion kill on the African savanna (although hyenas and jackals are also predators in their own right). In the oceans, remoras and pilot fish are scavengers that travel with sharks, eating fragments of the predator's victims.

Size is obviously an important element in food chains. For the most part, the predatory animals are larger than their prey—they have to be in order to catch and devour it. But there are exceptions. For instance, short-tailed weasels (known as stoats in Britain) will kill and eat rabbits several times their size. Generally, however, there is a limit to the efficient size of a predator. To support themselves, lions and sharks must range over a wide area. Even the densest population of lions in Africa, in Lake Manyara National Park, has only one lion per square mile on average. If there was an animal adapted to preying on lions it would have to be very large and powerful, and fast-moving, and it would need to range over an enormous area to support itself. So there is an upper size limit in food chains, and rarely more than six links are present between primary producers and top carnivore. The only consumers that can be huge and still be efficient are those that feed on thickly spread food sources (plants on land, plants and animals in the ocean plankton) and therefore do not have the high energy consumption of the carnivorous hunter. Two such groups of animals, for instance, are the elephants and the whalebone whales.

Food chains may be portrayed diagrammatically by pyramids divided into several levels, each representing a functional link in the chain: primary consumer, secondary consumer, tertiary consumer, and so on. These are known as *trophic levels*, each dependent on the one beneath it. Where the producers are small (as they are in the plankton and in grasslands) the relative sizes of individuals in the different trophic levels can be represented by an inverted pyramid with the pro-

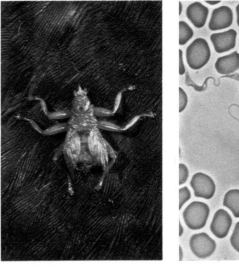

ducers at the bottom and the large tertiary, or higher, consumers at the top. The number of individuals at each level forms an opposite series: there are very many producers but very few tertiary consumers. Where the producers are large and few (such as trees in a forest) the base of the pyramid is distorted.

The trophic levels may also be compared according to their weight or *biomass*. A biomass pyramid is similar to a pyramid of numbers, the total weight of primary producers exceeding that of primary consumers, and so on. Such pyramids can be confusing, however, especially in the open ocean. The total weight of floating plants may be lower than that of the animals that feed on them, since these plants can be very short-lived and are constantly browsed by animals. They are growing and producing at a very rapid rate, providing sustenance for a larger weight of longer-lived and less frequently reproducing animals.

Further exceptions to the normal relationships in pyramids of size and numbers are presented by

A Predator and its Prey

- **Wildebeest**
- **Thomson's gazelle**
- **Burchell's zebra**
- **Lion**

Above: lions, which live in prides of up to four adults, must range over a huge area to get sufficient food. The diagram represents the Serengeti area of Tanzania, where there is an average of one lion to every five square miles—the area needed, roughly, to support some 350 individuals of its major prey species: wildebeest, Thomson's gazelle, and Burchell's zebra.

Below: the food pyramid shows that one square mile of the Serengeti area supports about 3 million pounds weight of plant life (green), but the plants support only some 25,000 pounds weight of large mammalian primary consumers (blue), such as wildebeest, Thomson's gazelle, impala, buffalo, topi, and giraffe, and these in turn support only 85 pounds weight of secondary mammalian consumers (red), predators such as lion, leopard, cheetah, hyena, and wild dog. Even so, this does not take into account the large numbers of small animals that also inhabit the Serengeti, but it gives a good example of the relationships in many land environments. One level of the pyramid must support the level above it on its excess production while maintaining a basic breeding stock, or it would rapidly disappear.

A Biomass Pyramid for the African Grasslands

Predators	85 pounds
Herbivores	25,000 pounds
Plants	3,000,000 pounds

parasitic animals and plants. These are organisms that live in or on the bodies of other living things, obtaining their food in a prepared state from the host. Unlike predators, parasites do not eat the whole body of the host, and they rarely kill it. The longer it survives, the more offspring will the parasites succeed in producing. Tapeworms absorb food products made digestible by the enzymes of the host's guts, fleas and leeches suck blood containing all necessary nutrient requirements and malaria parasites actually circulate within the bloodstream, absorbing food material. The broomrape plant gathers nutrients from its host plant's roots. A parasite's host expends energy in obtaining food for both itself and the parasite, so the host is the loser in a one-sided relationship.

Most wild animals and many plants are parasitized. The parasitic part of a food chain is an inverse pyramid—parasites are smaller and often more abundant than the host on which they feed. This food chain may also have several links. A jackal scavenging from a lion's kill may carry fleas that are themselves parasitized by protozoans. In some ways, parasites are very similar to scavengers, which feed on food obtained by the hunter, although they do not exploit the energy expended by the hunter in digesting food, as do many parasites.

Man is involved in a great many food chains at several trophic levels. He acts as a primary consumer of land plants, as a secondary consumer of herbivores, such as hoofed mammals, and as a top carnivore of fish. He harbors parasites and sustains scavengers. Many of the other organisms with which he is involved are members of still other food chains. And so we see how closely each food chain links with others. What we have, in fact, is not a number of separate food chains but a complex food web.

Matter and Energy: Circulation and Flow

The muscles of the cheetah bounding toward its prey and of the shark tearing at its victim are formed from material originally gathered by plants from the air, soil, and water. The fuel that powers muscle is also derived from plants. Through food chains, the matter of the earth's crust becomes rearranged into complex forms.

This matter does not accumulate at the top of each food chain, however. It is constantly being disarranged and returned to lower links; if it

The supple, bounding stride that can carry the cheetah at bursts of speed of over 60 miles an hour in pursuit of a gazelle is produced by muscles that have been built and fueled from the flesh of its plant-eating prey.

were not, the chains would cease to function for lack of material. This recirculation takes several different routes. Plants and animals excrete the wastes of their bodily processes and dump used material into their surroundings, and some substances diffuse out passively from their bodies. When they die, their dead bodies are consumed, through decomposition, or by other living things, releasing the building materials. The beauty and strength of living things depend on decay.

Bacteria are the chief agents of decay. Their rotting activities are of crucial significance in the circulation of the matter that passes through food chains. In fact, they occupy a distinct trophic level, separate from producers and consumers. Together with many fungi, they are classed as decomposers. Bacteria and fungi are commonly reckoned to be plants; but unlike green plants most of them cannot make their own food material by photosynthesis and must rely

on food produced by others. Some are parasites, using food from other living organisms, but the decomposers attack dead matter. In their feeding they secrete enzymes that break down the complex organic matter and release many of its contents so that they are available again at the base of the food chain.

Decomposers are most abundant, therefore, wherever wastes and dead matter collect, whether in the soil (inhabited by both bacteria and fungi) or in the surface and bed of the sea (inhabited by bacteria only). Bacteria are so small and so numerous that there may be as many as 455,000 million in a pound of soil and in some waters 570,000 million to a pint. The food of these myriads of decomposers forms the basis for a whole set of further food chains, much as carbon dioxide and water do for green plants.

Bacteria and fungi are consumed by a variety of animals. Bacteria form the major food of

Below: burying beetles interring a dead mole. The burying beetles and the fungi (bottom) are part of the mostly invisible community of decomposing and scavenging organisms that thrive on and under soil surfaces throughout the world.

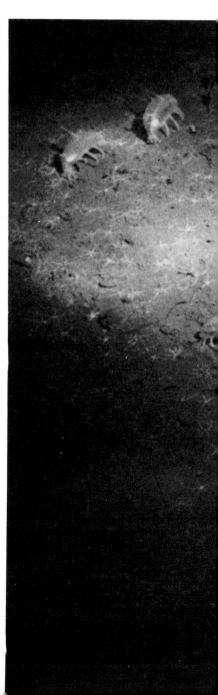

Below: fruiting bodies of fungi sprouting from elephant dung.

countless mites in the soil, as well as of many single-celled animals (protozoans) in the water and soil. Larger animals (man included) feed on the fruiting bodies of mushrooms and other fungi. Most of a mushroom's "body," incidentally, is a vast network of tiny filaments spreading through the soil, digesting dead plant matter. Not all the feeders on waste and dead matter are bacteria and fungi, however. An important part of the process of decomposition is the breaking down and mixing of waste and dead matter by the so-called *litter animals.*

In the soil lives a whole community based on the rain of material falling from the food chains above. Dead vegetation is avidly consumed by earthworms, millipedes, ants, and fly larvae. These litter animals are especially dense in temperate woodlands, where a thick layer of humus forms during the leaf-fall. (In tropical countries earthworms are less common. Their role is taken

The sea, too, has a large community of scavengers. The brittle-stars and sea cucumbers pictured here illuminated by the lights of a bathyscaphe at a depth of 3500 feet, feed on the rich sea-floor mud, extracting bacteria and other organic materials that result from a rain of dead organisms and particles of detritus from the surface.

Nitrogen gas in atmosphere

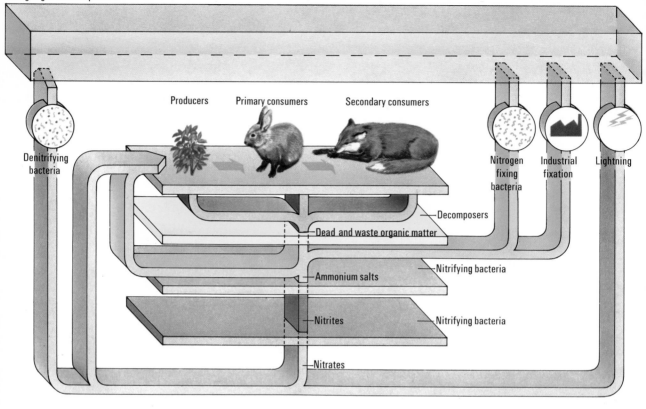

Producers

Primary consumers

Secondary consumers

Denitrifying bacteria

Nitrogen fixing bacteria

Industrial fixation

Lightning

Decomposers

Dead and waste organic matter

Nitrifying bacteria

Ammonium salts

Nitrites

Nitrifying bacteria

Nitrates

The Nitrogen Cycle

Nitrogen cannot be used directly by animals and green plants. Green plants, however, can take in nitrogen as dissolved nitrates and ammonium salts through their roots, and animals can obtain their proteins from these plants. This diagram shows the major ways in which atmospheric nitrogen gas (large red block at top) is fixed, that is, converted into ammonium salts (green) and into nitrates (blue). Natural communities rely mostly on fixation by bacteria. On land many nitrogen-fixing bacteria live inside the roots of particular plants, such as clovers. Most commercial agriculture, however, depends on nitrogen-rich fertilizers fixed by industrial processes. When plants and animals die or release wastes (brown), bacteria in the soil and on the seabed (nitrifying bacteria) break down the proteins and other nitrogen compounds in these remains, releasing ammonium salts. Then another variety of bacteria converts ammonium compounds to nitrites, and still others convert nitrites to nitrates. One group of bacteria (the denitrifying bacteria in the blue column on the left) inhabit the mud of swamps and other places where oxygen is scarce, converting nitrates back into nitrogen gas.

over by the far more numerous termites. Some termites live entirely on dead wood, relying on a rich population of protozoans and bacteria inside their guts to digest the wood for them.) Further up the food chain, predators such as carabid beetles, toads, and moles devour worms, and woodpeckers pry various wood-eating insects from dead tree trunks.

The bottom of the deep sea has its own "soil fauna," with animals and bacteria relying for food on the shower of dead material from the lighter regions above. One group of wormlike creatures that live in the mud of the seabed, the pogonophorans, seem to derive all their nourishment from the absorption of molecules that have escaped from the dead or living bodies of other organisms into the water.

When living things die and their bodies are attacked by bacteria and fungi, water is released from their tissues into their surroundings. Water, of course, is the basic constituent of animal blood and plant sap—the fluid in which the contents of living cells are dissolved. The human body is about 60 per cent water, many plants are 90 per cent water, and some water-living animals, such as jellyfish, contain even more. Water is constantly entering and leaving living things. It is lost through evaporation from the respiratory surfaces of land plants (in their leaves) and animals (through their gills and lungs), and is often used as the medium in which waste products are voided from the body. Water-living organisms usually exchange water continuously with their surroundings; and some inhabitants of fresh water even have to pump out the excessive amounts that flow into their bodies.

Water is made up of the elements hydrogen and oxygen. Together with carbon, these are by far the most common elements in living matter. Other important "raw materials" in the bodies of living things are nitrogen, sulfur, and mineral salts. These materials circulate constantly among plants, animals, the land, sea, and air. Green plants, as we saw in the last chapter, take in carbon and oxygen as carbon dioxide and synthesize this with water to form carbohydrates in the process called photosynthesis. From carbohydrates can be made fats and proteins, and these three substances are the basic materials for life on our planet. In addition, all proteins contain nitrogen, and some also contain sulfur.

To fuel their body processes (self-maintenance, growth, and reproduction), plants and animals take in oxygen from their surroundings. This oxygen has been released by plants from water during photosynthesis. It is used to oxidize carbohydrates, thus releasing energy—just the reverse of photosynthesis. This process, in which energy is released and made available for growth and other activities, is called *respiration*. During respiration every living creature produces carbon dioxide. This must be excreted from the body, usually by the same route through which oxygen enters—that is, through leaves, gills, or lungs. (In small water-living animals and plants, carbon dioxide may, like oxygen, diffuse straight through the outer surface.)

Most of the carbon incorporated into, and retained in, the tissues of plants and animals is eventually returned to circulation when the tissues are broken down and consumed by decomposers. The decomposers themselves respire, releasing carbon dioxide to their surroundings.

Respiration makes carbon dioxide swiftly re-available for photosynthesis. But not all carbon dioxide circulates through the biosphere quite so rapidly as in the photosynthesis-respiration cycle. Many living things build shells of calcium carbonate (made of calcium, carbon, and oxygen); and when they shed them as they grow, or when they die, these shells may accumulate on the seabed and build up into rocks. The famous white cliffs of Dover in England are made of chalk, formed on the seabed during the Cretaceous period (135 to 65 million years ago) from the thin shells coating certain minute planktonic animals called *Foraminifera*. Limestone is usually produced by the accumulation of animal shells. Some carbon in dead plant and animal bodies becomes trapped and fossilized before the bodies completely decompose. Peat, coal, and oil were formed in this way. This removal of carbon from the photosynthesis-respiration cycle is to some extent made up for by the carbon dioxide that is released from rocks by volcanic eruptions and by erosion through weathering and, today, by man's burning of fossil fuels.

The fourth most important element found in living matter—after hydrogen, oxygen, and carbon—is nitrogen, the most abundant gas in the atmosphere. Though some atmospheric nitrogen enters the soil as a result of lightning discharges, most of it is made available to food chains by the activities of nitrogen-fixing bacteria. These convert the gas into nitrates and ammonium compounds that green plants can absorb. Nitrogen is returned into circulation when the proteins of dead plants and animals disintegrate, and when animals excrete urea or uric acid after breaking down proteins in their food. Complex bacterial action eventually converts the excreted matter into nitrates. Some of these nitrates are taken up by green plants, as are the ammonium salts produced during the conversion process, but others are broken down by denitrifying bacteria, releasing nitrogen back into the atmosphere.

Some matter, such as mineral salts, is returned directly to food chains when the organism containing it dies and is decomposed. Plants living at the place of excretion or decomposition absorb the raw materials made available. But not all this matter is trapped by plants. On land, many soluble minerals are washed through the soil and into rivers by rainfall. This, known as *leaching*, is especially pronounced in wet tropical conditions, where decay is rapid and rainfall may be

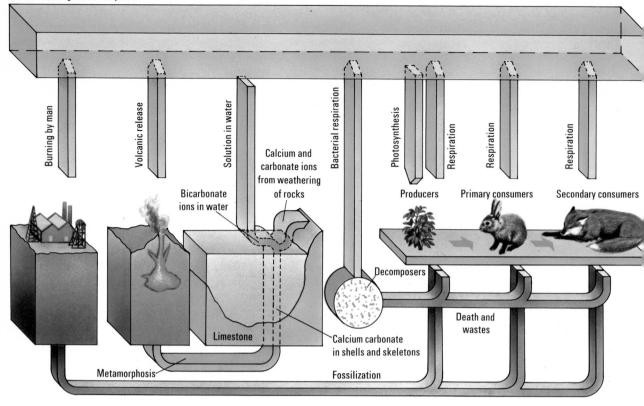

Carbon dioxide gas in atmosphere

Burning by man

Volcanic release

Solution in water

Bicarbonate ions in water

Calcium and carbonate ions from weathering of rocks

Bacterial respiration

Photosynthesis

Respiration

Respiration

Respiration

Producers

Primary consumers

Secondary consumers

Decomposers

Death and wastes

Limestone

Calcium carbonate in shells and skeletons

Metamorphosis

Fossilization

The Carbon Cycle

The diagram shows the constant exchange of carbon dioxide between the rocks, waters, and living things on the earth's surface. Land plants take in carbon dioxide gas (green block at top of diagram) during photosynthesis, when it is joined with water using light energy to produce carbohydrates. Carbohydrates pass along food chains to animal consumers (represented by the rabbit and fox in diagram). When living things use carbohydrates during respiration to yield energy, carbon dioxide is released back into the atmosphere. When living things die or produce wastes, carbohydrates in these remains are broken down by bacteria (decomposers). Sometimes the remains become fossilized to form peat, coal, or oil, from which man may release carbon dioxide by burning. Aquatic plants get their carbon dioxide from water in which it is dissolved as bicarbonate ions. Many sea creatures combine bicarbonate ions with calcium to form shells or skeletons, which after death accumulate on the seabed and form rocks such as limestone. The weathering of rocks and volcanic eruptions constantly supply carbon dioxide to the earth's waters and atmosphere.

sudden and heavy. In temperate climates, rotting proceeds more slowly and rainfall tends to be more moderate.

The rainfall that brings about leaching is by no means an entirely harmful process to land-living plants and animals. Apart from the fact that the water is essential for their existence, it is also largely responsible for making important minerals such as potassium, calcium, and phosphorus available in the first place. Water releases such

materials by dissolving the solid rocks through a process known as weathering. Weathering may be speeded up by the products of bacterial decay. It produces the soil, the mixture of broken-down rock and bits of dead organisms that coats most of the land in a thin layer. In dry areas, minerals dissolved in water are carried upward through the soil by surface evaporation. But in places where the climate is wet, the downward process of leaching often occurs at a greater rate than the upward flow. That is why land that has been cleared of tropical rain forest will not support much agriculture. The upper layers of soil under the forest contain few soluble minerals, for they are rapidly washed downward. However, over hundreds of years the vegetation itself has built up a big store of minerals, which have circulated through the living system with little loss. If the vegetation is cut down and removed, the minerals go with it. If it is burned on the spot, some of the minerals are blown away in the ash, and many of those that remain are quickly leached from a soil that has lost its complex living net of roots.

Water is constantly carrying matter from the land toward the sea, and in the sea there is a steady fall of matter to the bottom. So raw materials that are needed by living things accumulate on the beds of the oceans (and of lakes). Even from here, though, they are recirculated. Over a short period some are brought back from the bottom in upwelling currents. Upwelling commonly occurs when cold water sinks and pushes up warmer water from the lower layers. Such a phenomenon is especially characteristic of polar waters, where the upwelling of mineral-rich currents supports a dense community of plankton and other marine life.

Some of the minerals, too, are borne back to the land from the sea by wind and rain. Over a longer period—millions of years—such ocean sediments as chalk, limestone, and sandstone are carried above the waterline to form new land. Much of this happens through sea-floor spreading, which is responsible for continental drift. When drifting continents collide, squashing out the sea between them, the sediments at their margins become folded and thrust up, often high above the original sea level. Changes in the level of the sea itself (which can be caused by fluctuations in the amount of water trapped as ice at the poles), together with the rise and fall of continental material, can also act to expose sedimentary rock above water.

Apart from the influx of a fine rain of meteorites, the amount of matter coating the earth remains fairly constant. But, as we have now seen, geological and climatic processes and the growth and activities of innumerable plants and animals cause the matter to be continually circulated and rearranged. This rearrangement requires a good deal of energy.

Almost all the energy used by living things in their rearrangement of matter is derived from the sun. Unlike matter, however, it is not the same energy that is being constantly used and reused. It flows onto the earth's surface as sunlight, and then flows through food chains, performing building work in plant and animal bodies. As this work proceeds, the energy becomes changed to heat. Heat is a random form of energy—in other words it is the energy possessed by molecules of a substance moving at random—and it tends to appear whenever one form of energy is converted into another. All the light energy that flows into food chains is eventually transformed into heat and radiated away from the planet. Without the constant influx of sun energy that the earth receives, life would not continue.

We measure energy flow and the utilization of energy in living systems in terms of calories, which are a measure of potential heat production. Every year, an amount of solar energy equivalent to 1275 million calories for every square yard of the earth's surface enters the earth's atmosphere. More than half of this energy is scattered by dust particles in the air or used up in the evaporation of water. Of the rest of the solar energy available to plants, over 95 per cent is usually either reflected or used up in evaporating the moisture that lies on the plant's exposed surfaces. Only in a few exceptional cases do plants use for photosynthesis as much as 20 per cent of the light falling on them. For instance, the energy received by a square yard of field vegetation in Michigan, USA, has been measured at 392,500,000 calories per year; and only 4,860,000 of these calories—an efficiency of only 1.2 per cent—were actually incorporated into photosynthetic production in one recent year.

The energy trapped by photosynthesis can be released by burning the plant material. In forest and grassland fires, vast amounts of solar energy are released, but in normal circumstances plants operate their own internal processes that, chemically speaking, are similar to burning. As we have seen, respiration in the cells of a living plant is the work process responsible for maintaining life systems, for growth, and for reproduction. It uses up a large percentage of the solar energy trapped by photosynthesis. In the Michigan field experiment, 733,000 calories were lost each year in respiration, leaving a net production by photosynthesis of only 4,125,000 calories available to the next part of the food chain. This is only about one per cent of the sun's energy available to the plants. We are dealing

here with production rather than the total plant material (or "standing crop") available to the next level in the food chain. If more than the net production is consumed, the source of production is being destroyed.

The energy used by plants in respiration is lost from their structure as heat. When the sun is not shining on them, functioning plants tend to be slightly warmer than their surroundings. In an attempt to keep some of this heat in, to avoid frost damage, and to work more efficiently, many plants in cold climates have evolved special insulating structures or growth forms. Such plants are usually compact in shape, thus exposing a

minimum amount of heat-losing surface in relation to their volume; and they sometimes have an insulating covering of hairlike structures.

Further amounts of energy are lost by the consumers as they devour plant production. Most consumers use up some energy in moving to find their food, or even—as in filter feeding—in driving a current of water toward them. This work releases further heat. We know from experience that our muscles warm up when they are used in movement. More energy is lost—and more heat is produced—as consumers respire to power their other bodily activities, to grow, and to reproduce.

White Island is an active volcano in the Bay of Plenty, off New Zealand's North Island. Like most other volcanoes it is associated with a zone where two great plates of the earth's crust have met, with one being pushed down into the mantle. These volcanoes and the plate movements are responsible for raising rocks above the sea surface to begin a new cycle of weathering.

Like plants in cold climates, many of the consumers have developed means of trapping some of the heat they produce. Birds and mammals, for instance, maintain a constant body temperature by holding in the heat of respiration with insulating feathers, hair, or fat. This enables them to remain active in conditions that slow other animals, such as amphibians and insects, almost to a standstill.

Not all the heat that allows animals to operate more effectively comes from respiration. Many, especially those land animals that cannot maintain a sufficiently high body temperature, gain heat by basking in sunlight. In so doing, they are tapping solar energy directly, without depending on plant production as the intermediary. However, they acquire only heat and not a more useful form of energy in this way.

The loss of heat at each level of a food chain, as well as the loss of energy involved in making the transfer from one level to the next, is a primary reason why there is usually a progressive decrease in the total weight of organisms at each upward level in a chain. The energy requirements of a secondary consumer are often greater than those of a primary consumer; but in respiration (including that involved in moving) primary consumers have used up a good deal of the

Energy Flow Through a Living Community

The flow of energy through a living community based on a European oak tree. Only about 5 per cent of the incoming solar·radiation is used by the tree in photosynthesis; the arrows in this diagram represent the subsequent losses and transfers of energy through feeding, respiration, death, food wastage, and excretion along food chains. Very little of the original sunlight falling on the community reaches tertiary consumers such as the hawk shown here. In temperate deciduous woodlands a large part of the energy taken up by the plants goes to the bacteria and fungi that decompose fallen trees. This leaf litter supports a second food chain, called the decomposer, or detritus, chain to distinguish it from the grazer chain based on the living parts of the green plant. All the energy trapped by a living community during photosynthesis is eventually radiated back into space (top right-hand corner).

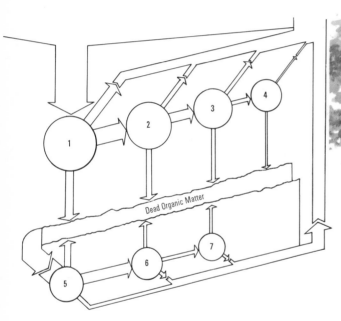

Dead Organic Matter

The major routes of energy loss and transfer are shown by different colors:

Yellow: Inflowing solar radiation (only 5 per cent used by plants)

Pale green: Feeding in grazer food chain

Dark green: Feeding in decomposer food chain

Red: Death, food wastage, and excretion

Blue: Respiration

The stages in the food chains are as follows:

Grazer Food Chain

1 Primary producers (oak leaves)
2 Primary consumers (e.g. caterpillars)
3 Secondary consumers (e.g. insectivorous birds)
4 Tertiary consumers (e.g. hawks)

Decomposer Food Chain

5 Decomposers (bacteria and fungi)
6 Feeders on fungus and detritus (e.g. springtails)
7 Soil predators (e.g. pseudoscorpions)

Energy is constantly exchanged between living things and their environment. Water condensing from the breath of this moose (left) on the Alaskan tundra gives striking evidence both of the mammal's high temperature, and of the loss of heat to its surroundings.

Bottom left: this iguanid lizard, basking to gain energy directly from the sun, displays a common reptilian habit. Reptiles are active when warm, but they cannot maintain a constant high body temperature independently of their surroundings, as birds and mammals can.

Below: a shaggy overcoat covering a fine, dense inner layer of wool insulates musk oxen's bodies against the bitter cold of the North American tundra, and helps to reduce heat loss.

plant's production of energy. The secondary consumers must then expend further energy—and so on right up the food chain. In any given area there is very much less energy available in the food eaten by top carnivores than was received originally by plants.

The flow and dissipation of energy does not always result in a lower weight of organisms as a food chain is ascended. We have already seen that planktonic ocean plants may be outweighed by planktonic animals. The point is that the *flow* of energy through the plants is always greater than the flow through their consumers, even if the actual amount trapped in the plants at any one time is less than that trapped in the animals.

A great deal of the energy that flows into primary producers eventually passes, like matter, to decomposers. These, too, produce heat in the respiration for which the complex energy-rich materials (from the waste products and dead bodies of other living things) are the fuel. The warmth commonly radiated from decaying matter is an indication of the high rate of bacterial respiration in progress.

Decomposers usually account for a larger part of the energy flow through a food chain than do the different levels of consumers, for decomposers act not only on consumer material but also on dead parts of producers—parts that have never reached the consumer level of the food chain. Although invisible to the naked eye, bacteria have a tremendously significant role in two fundamental ecological processes: the flow of energy and the circulation of matter. Large animals and plants seem important to man, but—as we have seen during the course of this chapter—their ecological significance is more than matched by that of the tiniest microorganisms.

A beech seedling thrusts up through the woodland litter. Intense competition for light, space, and nutrients means that only a very few tree seedlings will grow to maturity.

the individuals that most quickly cover the greatest area will best exploit the available resources—at the expense of slower growths.

This kind of competition occurs between members of the same species (*intraspecific* competition) and between members of different species (*interspecific* competition). As we have seen, it may involve speed of growth and the ability to exploit extreme conditions, but some forms of competition—among plants as among animals—may be more direct. Many plants, for instance, produce root secretions that inhibit the growth of other species, or even that of other members of their own species.

As the tangle of growth covers the forest clearing and the scrambling plants flower and seed, they create new shade beneath them, and this discourages more of the secondary growth. Seedlings of small trees spring up and reach rapidly toward the light. They soon cut off resources from the original colonizers, which may die (but the descendants of which may have already helped to keep the species going by springing up on the grave of another fallen forest giant). These small, fast-growing trees are themselves replaced by a slower-growing species, which have more "shade-tolerant" seedlings that dominate the mature forest. These compete as they grow up through the smaller trees, cutting off resources from them in their turn, until eventually only one or two large trees dominate the others and replace the original inhabitant. The forest has now reverted to its previous appearance.

This type of plant *succession* is a good example of competition for resources among primary producers. Green plants are engaged in a continual silent and subtle struggle for resources, reaching for their nourishment upward, downward, and sideways into the limits of useful space wherever conditions are suitable. And in order to reproduce, those that rely on animals to fertilize their flowers attract the pollinators by a variety of smells or colors, or even by flowering when few of their "competitors" are doing the same.

Just as the primary producers compete in colonizing cleared ground, so do many decomposers compete when they invade a dead body—the equivalent to them of a piece of virgin soil to plants. Spores from a host of bacteria and fungi soon alight on dead plant material, just as seeds fall in a clearing. The spores germinate, the fungi spread their threadlike feeding hyphae through the food source, and the bacteria spread

Young sycamores reach toward the light pouring into a clearing in a European beechwood. The sycamore, a fast-growing tree that springs up on cleared ground, is a typical colonizing species. In the course of time, as the plant succession progresses, beech trees are likely to replace these sycamores. Below the sycamores ferns are being slowly excluded from the light.

animals off plant and stone surfaces with its wide mouth. The kululu also feeds in shallows, but concentrates on sandy areas. It picks up mouthfuls of sand, swishes the sand around its mouth to separate organic material from the grains, then swallows the organic material and spits out the sand. Such examples of similar species that avoid direct competition by occupying slightly different niches in the same habitat are common all over the world.

Direct competition may lead to the extinction of a species. This often happens—and can be seen happening—as a consequence of man's introducing species into parts of the world where they do not naturally belong. Even if carefully controlled at first, these artificially introduced species frequently escape into the wild. If they thrive, it is often only at the expense of native populations. Island faunas are particularly susceptible to this form of competition. By natural means, only a limited number of land animals reach islands; there, cut off from the "mainstream" of evolution, they often develop in unusual ways, adapting to the island conditions in the absence of competitors. When man introduces species from other parts of the world, the native animals may not be able to compete for the island's limited resources. Classic examples of this process have occurred in the Galápagos and Hawaiian archipelagos, in the West Indies, and in Australasia.

In the Galápagos, for instance, the chief large vegetation browsers used to be giant land turtles, but these have been brought to the verge of extinction on many of the islands through competition with non-native goats. The goats are more efficient browsers, and so the turtles have been increasingly unable to find enough food left in places they can reach. In Australia, cats, foxes, goats, and rabbits have severely affected the native pouched mammals, the marsupials.

It is not only the species *introduced* by man and now running wild that provide striking examples of interspecific competition. Over much of the world, *man himself*, with the domestic plants and animals he raises, is the greatest competitor for space, materials, and energy.

Nor, as we have said, is competition limited to the kind that occurs between species; it also occurs between members of the same species. As with interspecific competition, the most obvious form of intraspecific competition is for space. As a secondary effect, this usually involves com-

petition for food, or mates, or both. *Balanus* barnacles, for instance, do not kill only *Chthamalus* barnacles. They compete with each other for the limited space available on the rocks of the shore, where space gives access to both food and mating partners. The *Balanus* population of Britain suffers a heavy death toll from being displaced, crushed, and smothered by members of its own species.

Territorial behavior is a common form of intraspecific competition among many backboned animals. The territory is a relatively fixed geographical area, which is defended against intrusion by other individuals, or social groups, of the same species. A pride of around a dozen lions (which would commonly contain two adult males, four to five females, and about five younger animals) will defend an area of African savanna against neighboring prides partly by loud roaring. In South America small groups of titi monkeys (which consist of a mated pair and their offspring) defend small areas of rain forest (about one acre) at the territorial boundaries by bristling their hair and lashing their tails. Territories are usually defended by display, rather than wasteful fighting, but chases and fights do sometimes take place.

Lions and titis have territories that remain theirs for long periods. Many animals, though, become territorial only during the breeding season. Often, the males of such species hold territories against other males, and the females enter the territories in order to mate with the holders. In European ponds, male three-spined sticklebacks assume gaudy coloration in the spring breeding season and simultaneously set up territories; the colors warn off other males who might stray into the area. The bright plumage of many male birds serves a similar function of chasing away intruders. For the same purpose, the red breasts of the unrelated European and American robins and the scarlet epaulets of the American red-winged blackbird are displayed in combination with song. The Old World warbler, more drab in color, announces the position of its domain by sound alone; each of the many species that live in temperate woodlands has its own particular territorial song.

An unusual form of competition takes place in dense populations of the antelope known as the Uganda kob. The kobs have all year round special breeding grounds of a roughly circular shape, some 500 yards in diameter. On these grounds the

Right: a small, spiny rayed male dragonet of European shores displays his bright blue breeding-season decorations on body and fins, which probably serve to attract females and to warn off other males.

Below: in competition for mates many adult male animals establish territories in the breeding season. Here male Uganda kob antelopes battle with interlocked horns for possession of a specially favored territory on which they can mate with females.

males compete for possession of about a dozen central, circular territories that are each only about 20 to 60 yards wide. Once a male has gained one of these territories, he may be able to hold it for only a few days, displaying with lowered ears and often fighting with his lyre-shaped horns. Competition is for mates, as females pair most often with the males that hold the central territories. Some birds, such as black grouse and the South American cock-of-the-rock, have a similar competition when they gather together in the breeding season.

Ecologists have been arguing for years about the significance of territorial behavior. One result of it may be to space animals through their habitat in such a way that the available resources are divided up adequately. Territory-holders ensure that they have necessary supplies for themselves or their groups, and those that cannot obtain a territory in favorable surroundings may starve, or fall victim to predators or disease. Territorial behavior may, therefore, help to control population size because there is a limit to the number of territories that can exist in any particular area, and it seems to be true that animals without territories rarely breed. Where territories are held and defended by male animals, the behavior acts as a form of natural

selection between them, for the most competitive animals will tend to father the most offspring.

Although individual plants and animals are engaged in a continual competition for resources, they must also, to some extent, cooperate with one another in order to be successful in this competition. The most obvious examples of cooperation are those between individual members of the same species, which help each other to survive and reproduce. Except where self-fertilization occurs, cooperation is necessary for sexual reproduction itself: the mixing of hereditary material from different individuals. This form of cooperation takes place in the life cycles of most types of plant and animal, even those that reproduce mainly by budding or splitting.

At a very basic level, cooperation in sexual reproduction involves structural matching, for the genetic material from the two parents must be able to mix. In other words the sex cells containing the genetic material must reach each other, and must then be capable of recognition and fusion. Many plants and animals do not actually meet as individuals for sexual reproduction, but release sex cells into their surroundings. These cells recognize their opposites by chemical means. For instance, the sperms of sea urchins have molecules on their surfaces that

Above: male and female wandering albatrosses dance on Bird Island in the subantarctic. Their prolonged courtship establishes a tight bond throughout the breeding season, to be reestablished in future years.

Left: southern right whales cavort in ponderous courtship chases before mating in a sheltered gulf on the Patagonian coast, where they gather each year. A year later a single calf will be born, and soon afterwards the female will mate again. These whales, formerly the mainstay of the whaling trade, are almost extinct.

65

Like all monkeys, these macaques of Southern Asia live in social groups; by grooming each other they not only remove parasites but also reinforce friendly relationships.

match the molecules on eggs released into the sea at the same time. And the pollen grains of a flowering plant (which are the male sex cells) begin to germinate when they land on the stigma of a flower belonging to the same species. If they meet a stigma of a different species, they usually fail to germinate, or grow only slowly.

Most freely moving animals meet for reproduction, and when they do a complicated courtship often takes place before mating. The dance of a pair of wandering albatrosses, the elaborate building and display of the male bowerbird or the male stickleback, the strutting of many buck antelopes, the vibrating song of male grasshoppers—such gestures invite the cooperation necessary for reproducing the species. One sex produces a sign that the other recognizes and responds to, ensuring that an individual will mate only with a member of its own species, and with one in a reproductively active condition.

Following courtship, the male and female may mate in one of two fashions. They may both deposit sex cells into their surroundings: the female stickleback, for instance, lays her eggs in the male's nest, and the male then deposits sperm over the eggs. Or, as commonly happens among insects, reptiles, birds, and mammals, the male introduces sperms directly into the female's reproductive tract.

As in all cooperative endeavors, good timing is important for sexual reproduction. In the species that do not meet in order to mate, cooperation must be achieved through remote control. This means that the sex cells in different individuals must be synchronized in their development. It would be useless for millions of pollen grains to be dispersed on the wind at a time when the ovaries of receptive flowers were not ready to receive them, or for the many sperms of sea urchins to swarm in the absence of eggs. The necessary remote-control timing is usually brought about by climate. Widely separated plants will flower together, triggered off by particular conditions of day length, temperature, or rainfall, or by a combination of these factors. Climatic factors bring sea urchins into reproductive conditions, and chemicals released when sperms are set free cause the discharge of eggs from other urchins. In the coral reefs of the Pacific and West Indies, the sexual cycle of palolo worms coincides with the phases of the moon, so that millions of writhing egg- and sperm-bearing segments are discharged into tropical waters in the early hours of a single morning.

Even the species whose individuals do meet will often rely on climatic factors as an external clock to bring their reproductive organs into the right condition at the right time for successful mating. The changing seasons, noted by an animal's nervous system, will influence hormone production. In turn, this affects the development of sperms and eggs in testes and ovaries. Thus mating can take place at the best possible time of year—the time, in fact, that will provide the most favorable conditions for the growth of the following generation.

The nurturing of the young is another aspect of reproduction in which cooperation between individuals is often of vital significance. Plants, as well as most simple animals, abandon their offspring to fend for themselves. Many parents, especially land-living insects and vertebrates, care for their young during the early period of growth, and so increase their chances of survival. Parent birds frequently cooperate closely for long periods throughout the breeding season—in building the nest, incubating the eggs, and feeding the young until they are able to fly. Such prolonged cooperation explains why bird courtship is often so involved. A strong bond must be forged between the pair, for it has to last from mating until the fledglings become independent.

The wandering albatross has a truly extensive courtship period. The young wandering albatrosses start to court on the subantarctic island of their hatching from the summer of their fourth to fifth year, but they do not actually breed until the age of about 10. Then, at least a month of

A pair of Caspian terns cooperate to rear their small brood. The intense care shown to their offspring by many animals increases the offspring's chances of survival.

intensive courtship and nest-building precedes the laying of a single huge egg, which the parents take turns incubating for nearly three months, with time off to gather food at sea. After the egg hatches, both feed the large, down-covered chick. When the chick is a month old it is large enough to be left alone while the parents forage for regular supplies of fish and squid for their rapidly growing offspring. The following summer, at the start of a new breeding season, the young, fully fledged albatross is ready to take wing over the great expanses of the southern oceans.

Typically, animals that exercise a great deal of parental care have few offspring. There are two effective and opposing reproductive extremes in the living world: energy may be expended either in the production of many sex cells or in caring for the product of the fusion of the cells. So, either vast numbers of young may be produced

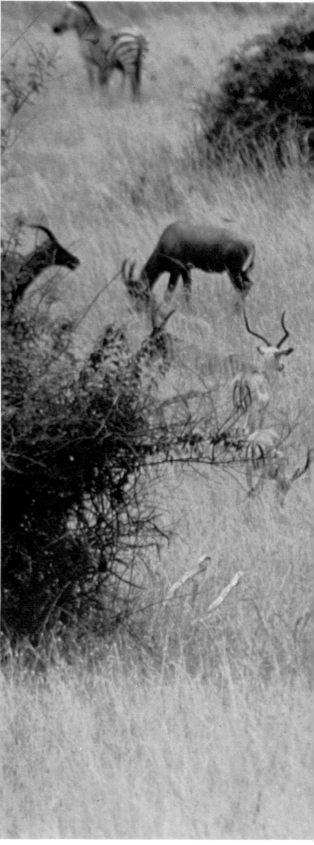

Different species of herd-living mammal, such as these Uganda kob (foreground), topi, and Burchell's zebra, are often found together on the African savanna. These associations may arise through the animals feeding on similar vegetation. Like the fish schools, they benefit from the association, as each species responds to the danger signals of the others.

Below: members of this school of fish benefit from their life together. Their numbers confuse predators and greatly multiply the sense organs that detect food and danger. Most fish schools are probably less durable than groups of birds and mammals.

Right: with its pollen baskets laden with golden pollen, the returning honeybee just right of center "dances" on the surface of a hive, indicating the location of its find to others by the particular pattern of the "dance's" movements.

Left: Indian langur monkeys live in groups of from five to over 120 animals, in which adult females outnumber males by as many as six to one. The group communicate with each other through a wide range of gestures and calls. The males excluded from these groups live alone or sometimes in small bands.

(which are given no care and of which only a tiny percentage reach maturity), or there may be very few offspring, which are given intensive care and have a high chance of survival. Between these extremes exists a whole spectrum of compromises.

There is an even better chance of offspring surviving if family cooperation extends beyond the breeding period to encompass the whole life-span, or some large part of it. Many animals spend most of their lives as part of a social group. Living in close proximity to other members of their own species, they interact constantly and in coordinated fashion, to the mutual benefit of all. The benefits can be numerous. For one thing, the group provides a multitude of sense organs, so that one individual can warn many others of danger, or can locate food for the whole group. Then, too, members of a group can act together to confuse or repulse predators, or can themselves cooperate in catching their prey. In a group, energy need not be wasted in seeking out distant mates; and several animals—not just the parents—can help to protect and nourish the young. Finally, the group can act as a store of information: various individuals remember different things about the world they live in, and the separate bits of information can be pieced together for the benefit of all.

Small invertebrate animals, such as crustaceans in the ocean plankton and worms in the forest soil, often live close together, but they rarely lead a cooperative social life. It is the insects and vertebrates, the animals that exhibit the most highly developed forms of parental care, that most often live in complex societies. Shoals of fish, flocks of birds, colonies of termites, and herds of mammals—these are examples of social groups. The link with parental care is probably more than coincidental, for the social group probably has its evolutionary origins in the close-knit breeding-season family.

Some groups, such as those of South American titi monkeys and Asian gibbons, consist of only a breeding family: a mated pair and their offspring. Other groups are much larger: a colony of African driver (or safari) ants may number hundreds of thousands. The numbers in a group and its geographical range tend to remain relatively stable over long periods of time, despite the changes in the actual individuals present. In most mammalian groups, deaths from predation and disease are normally replaced by births. Usually, too, there is a slow rate of emigration and immigration; young males may be forced out of the group before reaching maturity by a dominant adult male.

The presence in some groups of a dominant male is an example of another common aspect of this form of cooperative existence: the division of labor. Different individuals in a group not only cooperate, but also frequently perform different roles, thus improving the group's efficiency. Obviously males and females perform different functions in the production of new generations— a difference that is particularly marked in mammals, where the female must protect and feed the young animal for a long period both before and after its birth. But the distinction of roles usually involves more than this, with different sexes and ages in the group all behaving in different ways. For example, an adult male monkey not only plays a different reproductive role from an adult female, but he is also usually much more concerned than she with group protection (watching for predators, warning of their approach, and helping to ward them off) and in

71

interactions with other groups (displaying territorial ownership and chasing away rival males). Meanwhile, young monkeys of both sexes spend a great deal of their time in exploration and play; in this way they get to know their fellow group members and the world in which they must live.

Division of labor has reached its highest development in colonies of social insects, such as termites, ants, bees, and wasps. Not only do the adults belong to different sexes, but different castes with distinct functions have evolved. The colonies usually live in a nest that "worker" members build for them—combs of bees, ant heaps, and termite mounds, for instance. (There are, as almost always, exceptions. Driver ants, for example, are nomadic, although they do excavate temporary bivouacs.) Ants, bees, and wasps belong to a group of insects known as the *Hymenoptera*. A colony usually contains a single reproductively active female and a few reproductively active males, together with large numbers of sterile females who form the worker caste. The workers serve the reproductive individuals by building and defending the nest, gathering food, and caring for the eggs and young. In ant colonies there may be many sub-castes of workers, specialized for a variety of tasks. Termites are unrelated to the *Hymenoptera*, but they have a very similar social organization, except that workers include sterile males as well as females.

A feature of insect societies is the complex form of communication that takes place between individuals. Termites communicate danger to one another by tapping their heads on the walls of

In some parts of Africa, piapiac birds—small crows shown here perching on the backs of some white rhinoceroses— obtain much of their food by eating insects disturbed by large mammals. If the piapiacs see approaching predators, their harsh alarm call will alert the mammal, so both species benefit from the association.

their nests, and they lay chemical trails for other members of the colony to follow. European honeybees are famous for their "dancing": when workers return to the colony after foraging, they move in a waggling fashion according to a specific pattern; the speed and form of this "dance" indicate the location of food sources they have discovered. Communication of some sort between individuals is necessary for social animals if cooperation is to be effective. All such animals have highly developed "languages," which use visual, vocal, or chemical signals, or a combination of them. Forest monkeys and apes communicate with a particularly wide range of signals. Indian langur monkeys, for instance, have a language consisting of at least 15 distinct calls, 15 gestures, and 9 different ways of touch-

ing each other. Used in combination, these signals allow the monkeys to exchange a great deal of information about their environment and their own state, helping to make the group an efficient functional unit in its interactions with its surroundings. Man himself—a relative of the monkeys and apes—is a social animal with well-developed vocal language, which has enabled him to develop a society of enormous complexity.

The development of our modern society has relied heavily on the cultivation and domestication of other species, plant and animal, from which we obtain food, clothing, building materials, and transportation. So we cooperate not only with other people, but with other species. Such cooperation is rather one-sided, however, for it has come into existence through our having grossly modified the way of life of wild forms. In natural communities there are many examples of a more nearly mutual cooperation between different species—relationships in which two or more species play active roles and all benefit from the association. Such cooperation may not involve a close physical contact. It is commonplace for different bird and mammal species to warn each other of danger, just as do members of a social group of the same species. A bushbuck or duiker antelope that detects a human hunter in the gloom of an African forest will bark an alarm call as it flees, and monkeys in the trees above will respond to this alarm and call out themselves, so warning others of the danger. In itself, this is more exploitation than cooperation, for the antelope does not gain from the monkey's response (although in the future it may itself be warned by a monkey-initiated alarm). However, association between monkeys and antelopes may go further than this. A red duiker may travel for long periods beneath a group of red-tailed monkeys, feeding on vegetation dropped from above as the

monkeys forage. In this way the duiker benefits from the monkeys' activities, and its presence on the ground gives the monkeys an extra sentinel, which can warn them of danger.

Outside the forest, the African savanna provides an example of a closer feeding-warning relationship between species. Oxpecker birds, which are related to starlings, perch on the backs of many large herbivores, especially buffaloes and rhinoceroses. An oxpecker on a rhino's back often has a clearer view than its host, whose vision (probably poor anyway) can be obscured by long grass and other vegetation. If the oxpecker spies approaching danger, it flies up and gives a buzzing alarm call, alerting its host. The oxpecker benefits from the relationship in two ways: it eats ticks and tsetse flies off the host's back, and it gains warmth at night from its close association with the big animal. And the mammal benefits not only from the early-warning mechanisms but also by being kept relatively free of parasites and biters.

Oxpeckers do not live permanently with a single herbivore. But there are some species of organism that spend their whole lives with another species in cooperative interaction. In an earlier chapter we discussed one instance of this type of association: the way some protozoans live permanently inside the guts of secondary consumers such as termites and ruminant mammals, helping them to digest their food and in return gaining a sheltered environment and food for themselves. Such mutually beneficial living together is called *symbiosis*.

Algae are especially common symbiotic organisms. A lichen consists of a fungus combined with a colony of symbiotic single-celled algae. The fungus provides the algae with protection, water, and minerals; in return, the algae produce food for the fungus by photosynthesis. Many marine animals have large numbers of brown algae, known as *zooxanthellae*, in their tissues. The algae occur in the fleshy tissue inside and along the rim of giant clam shells, in corals, in sea anemones, and in many other invertebrate animals. Although the animals do gather food for themselves, they also get large quantities from the photosynthesis of the algae, which in turn gain protection and rich supplies of carbon dioxide from their hosts. The algae's removal of carbon dioxide from reef-building corals may also speed up the rate at which the corals can produce their calcium carbonate reefs.

Through cooperating closely with one another in such ways, species can increase their efficiency in the competitive struggle for the space, energy, and materials that they require for growth and reproduction.

The plantlike corals, such as the branching corals pictured here with brain corals in the background, are actually colonies of animals with photosynthetic algae living in their tissues. The large tubular forms (center) growing on the reef are sponges.

Ecosystems: Life and Environment

The trailing tentacles of a tiny hydra, hanging from the underside of a lily pad in the richly vegetated waters of a pond in the northern temperate zone, ensnare a passing water flea. The hydra is green, for its tissue carries the cells of the symbiotic alga *Chlorella*—cells that provide a good deal of their host's food requirements. The water flea itself has browsed on the minute floating plants of the phytoplankton. These have used the energy from sunlight falling on the pond to build up their tissue from water and dissolved carbon dioxide. The phytoplankton has also relied on dissolved mineral salts brought in originally by a stream or by some other drainage water off the land. The salts are circulated in the pond by currents flowing up from the bottom, where additional minerals are released from decaying organic material. Without the inflow of water the pond would disappear evaporated by the sun's rays.

It is summer, and the hydra is by no means the only predator of water fleas. It must compete for its food with small fish, such as the stickleback, and with insect larvae and other miniature hunters. As winter approaches, the productivity of the phytoplankton and other vegetation declines, and with it the number of planktonic crustaceans. Many individual hydras die during the winter, but not before reproducing sexually. The thick-walled eggs that result from the sexual process survive the winter cold in the muddy floor of the pond.

This is a microcosm of events in the living community that extends over most of the earth's surface. Living things are mixed together, involved

A European pond as an ecosystem, showing some of its fundamental components: (1) energy flow, with floating and rooted plants trapping sunlight; (2) food chains, exemplified by a predatory fish eating an herbivorous snail; (3) decomposition and the circulation of matter, in which bacteria play a primary role; (4) cooperation between living things, such as the green algae living in the tissues of a hydra; (5) competition, shown by the merganser and otter, which both prey on fish. Arrows in the diagram indicate the circulating currents that carry material around the pond (brown), and the exchange of gases involved in plant and animal respiration between air and water (blue). The effects of the surrounding environment (windborne dust, leaves, twigs and other matter, drainage from land, and weather) influence all parts of the ecosystem.

with each other in a great variety of ways, including feeding, competition, and cooperation; energy flows through the community, and the physical environment has profound effects on all plants and animals. The environment of any living thing consists both of other living things and of inanimate materials. The living community and its nonliving environment act together as an ecological system—an *ecosystem.*

The whole surface of the earth can be regarded as an ecosystem—a functional ecological unit—that we often refer to as the *biosphere.* The hydra's pond is an ecosystem, too. It has particular physical characteristics and is influenced by particular weather conditions, which affect the nature and way of life of the plants and animals found within it. It can be studied as a functional unit within which we can investigate food chains, pyramids of numbers, energy flow, the circulation of materials, and so on. It is therefore necessary to set some boundary within which to study these various ecological attributes.

The boundaries of an ecosystem are not absolute, of course. They are usually defined by convenience. A pond—which is a small piece of water surrounded by land—is fairly well delimited. It cannot exist in isolation from the surrounding air and soil, but it can be studied, in a practical way, as a unit. The surface of the earth, on the other hand, is much less easy to study as a single entity. Ecosystems become progressively easier to describe and dissect as they become smaller. Africa can be examined as a unit more easily than the whole earth, the African tropical forest more easily than the whole continent, and one forested hillside more easily than the entire forest. But all ecosystems share certain common features. Each one of them is a dynamic mixture of the nonliving environment (of solids, liquids, and gases) with living things (producers, consumers, and decomposers), interlocked with great complexity.

Even the twig of a tree in the African forest can be regarded as an ecosystem. On the surface of the twig, which is surrounded by the gases of the air, rain falls and lichens may grow. These will be browsed by a variety of minute invertebrate animals, which may be eaten by larger animals that move among many twigs. An ecosystem is never totally isolated. The twig obviously is not isolated, nor is the forest—just as the earth is not alone in space; it is a satellite of the sun, which is itself but one star in a galaxy.

When the activities of plants and animals and their interactions with one another and the surroundings are brought together and examined as a functional unit, the importance of the physical, nonliving part becomes readily apparent. The ecosystem represented by the African tropical forest at the equator is totally different from that represented by Lake Victoria, by one of the Galápagos Islands, or by the open Atlantic at a similar latitude, just as it differs from the ecosystem of the North American Arctic, the Australian desert, or the Himalayas. These areas all have producers and consumers and other attributes of an ecological system, but the physical factors in each case are not all alike. Important among these physical factors are the relative amounts of water and land and their chemical composition, the amount of sunshine and annual variation in length of day, the movement of air masses (which particularly affect the amount of rainfall and its annual distribution), and the altitude.

These factors, which can be summarized as substrate (the physical foundation on which everything else rests) and climate, result inevitably from the nature of our planet: earth is a sphere spinning on its axis as it revolves around the sun, its surface covered unevenly with liquids and solids, and enveloped in a thin layer of gas. Variations of substrate and climate have produced a set of large-scale ecological systems over the earth's surface, and each of the systems has particular characteristics that mark it off from other systems.

Different ecologists recognize different numbers and kinds of large-scale ecosystems, but there is general agreement about some major types. Major types on land (sometimes called *biomes*) are tundra, northern coniferous forest, deciduous woodland, sclerophyllous woodland (containing trees with small, hardened, drought-resistant leaves), temperate rain forest, tropical rain forest, savanna, steppe, and desert. Major aquatic ecosystem types are lake and pond, river and stream, open ocean, seashore, and estuary. We need not examine every one of these major

This view of earth from space shows clearly some of the interactions of land and water that produce different major ecosystems. The clouds over Europe, central Africa, northern Asia, and the Indian peninsula indicate where winds are carrying the large quantities of water evaporated from the oceans by solar radiation. The pattern of atmospheric circulation supplies little water to northern Africa, the Middle East, and southwest Africa, and great tracts of desert result.

ecosystems in detail; for our purposes, a look at a selected few areas can serve to show the importance of climate and substrate in determining the structure and way of life of the local plants and animals.

A warm desert ecosystem is a striking example of the influence that the physical factors of the environment exert on a living community. In the moist, temperate regions of Europe and North America the impact of particular weather conditions on the ecosystem is not immediately apparent, at least in summer, to a nonecologist. In most deserts, on the other hand, the dryness and heat are severe at most times of the year, and the effect of such a climate on the plants and animals can be easily recognized.

Warm deserts exist in tropical and subtropical land areas that get very little rainfall because of the distribution and movement of atmospheric air masses. Life on our planet depends on water, which, as we have seen, forms the major part of all plant and animal bodies. Life is absent where water is absent, and scarce where water is in short supply. In deserts the density and biomass of plants and animals are much lower than in wetter areas at the same latitude, and plant production is similarly reduced. Deserts, in fact, can be more barren than the tundras of polar regions: they often produce less than one fiftieth of an ounce of organic material a day on each square yard. Extreme deserts may produce less than one third of an ounce in a whole year on each square yard.

Lack of water is the primary factor that has produced most warm deserts, although very salty soils in only moderately dry conditions can have a similar effect on vegetation. Desert generally results when there are less than six inches of rainfall a year in an area of warm climate. Between six and 16 inches of rain, the result is likely to be semidesert. When the rain does come in such regions, it is usually concentrated into a short period, so that a brief favorable spell for plant growth is followed by prolonged conditions of extreme unfavorability. Heat and the lack of clouds cause rapid evaporation of any rain that falls, and this accentuates the arid climate. Much of the rain that does fall may run off so fast that very little can be used by the plants. Scarcity of clouds, of dense vegetation, and of moist soil allows surface temperatures to rise scorchingly high. In some places the thermometer may register up to 185°F in the middle of

the day. More significantly, the difference between day and night temperature is usually great, commonly about 35°F, for the desert surface cools rapidly at night due to lack of insulation.

Despite the unfavorable conditions for life, many plants and animals have become adapted to the extreme demands of desert life. The cactus, a typical desert plant of the Americas, is well adapted to the water shortage. Some cacti have root systems that spread over vast areas near the surface of the soil—the root system of the saguaro cactus may have a diameter of 90 feet—ready to trap the maximum amount of moisture possible from the infrequent downpours. They store the trapped moisture in their fleshy stems; and evaporation is reduced by several characteristic features of the cactus: the absence of leaves (except for those modified as spines, to keep browsers from the precious water stores), the small amount of branching, the presence of a thick waterproof covering, and the prolonged closure of the stomata, which open only when the surrounding air is relatively moist to allow the passage of respiratory gases.

Following a rainy spell, a huge saguaro cactus may weigh as much as 10 tons, 9 of which are stored water. This water economy helps to conserve what little is available, but it means that the plants are inevitably much less efficient at photosynthesis than is a typical broad-leaved tree. Other desert *succulents* (the name given to water-storing plants such as the cacti) have similar adaptations.

Some plants have different adaptations for survival. African acacias and the American mesquite, for example, push their roots so deep into the soil that they can draw water up from very deep down. The roots of the small mesquite tree may reach as far as 50 feet below the soil surface. Some other types of desert shrub are deciduous: they shed their leaves completely during the driest part of the year, so losing less water at that time, but they retain a green stem that slowly carries on photosynthesis.

Many desert plants disappear completely, except for their seeds, during the driest weather. They are what botanists speak of as the *ephemerals*—plants with a very short growth period, which spring up after adequate rainfall, flower, and quickly produce drought-resistant seeds. The seeds survive after the plant itself has died, and germinate when favorable conditions recur. The ephemerals are always small plants, for an

Lack of water leaves some desert areas almost devoid of life. Without an abundance of plants, particularly grasses, to bind the soil, winds have blown these Saharan sands into ridged dunes. Windblown sand erodes the rocks to form yet more sand.

81

obvious reason: they are too short-lived to be able to grow big. Many are grasses, but others are herbs that produce especially bright flowers to attract insects, for rapid pollination is essential. *Boerhavia repens* of the Sahara, a relative of the South American *Bougainvillea*, is said to sprout, flower, and seed in only 8–10 days.

The coating of the seeds of many ephemeral plants contains special chemicals that prevent germination in dry weather. Even a shower or two will not affect such an inhibitor. A reasonably prolonged rain, however, washes the inhibitor out of the seed-coat, allowing the seed to sprout. Without such an inhibitor ephemeral seeds might germinate at a time when there would not be enough water to permit the full growth cycle from sprouting to fruiting.

Most of the consumers that feed on the desert plants are small and feed only at night, avoiding the heat and the dangerously high drying potential of the day air. Small rodents, such as jerboas and kangaroo rats, are particularly important as herbivores in deserts. They do not need to drink but can survive entirely on the water taken in with their plant food. This is possible partly because

of a special arrangement of blood vessels in the nose of such rodents that acts as a heat-exchanger (cutting down water loss during respiration) and partly through minimizing water loss by production of a very concentrated urine.

Many herbivorous desert insects (as well as crustaceans and snails) behave like the ephemeral plants in that they are active for only the part of the year when vegetation abounds. During the long, hot, dry seasons they survive as eggs or as dormant pupae or adults. Others, like the rodents, are active mainly by night. Some, such as carabid beetles, occupy the next level in the food chain, by preying on other forms of animal life. Among the desert predators we also find scorpions, and many different lizards and snakes. The waterproof external coverings and dry excretions of these animals make them well-suited to life in the desert. Most of the desert reptiles, whose body temperature closely corresponds to that of their surroundings, are active only in the early morning and at night. During the hot daytime hours they hide under rocks, in sand or among vegetation, or in their burrows.

This is only a brief glance at the desert ecosys-

Right: oryx antelopes of African deserts can survive without water during periods of drought by relying on the moisture obtained from their food.

Far right: the Australian Gould's sand goanna, or monitor lizard, is shading itself under a desert plant, holding its body off the ground to reduce heat inflow.

tem, in which the web of relationships may be just as complex as that of many other environments. But although the roles occupied by desert organisms, and their interactions with one another, are similar to those found in quite different ecosystems, the particular form and behavior of the organisms and the amount of solar energy that the system as a whole uses are markedly affected by its special nature. The desert's physical conditions—most notably an extremely arid climate —can support only such a living community as we find existing within its boundaries.

The influence of climate on ecosystem can be seen vividly when we journey from the desert to a very different kind of place. Traveling south from a desolate wind-blown *erg* (a desert region of shifting sand dunes) near Agadez in the southern Sahara, at a latitude of 15° North, we find ourselves after only about 300 miles at Kano in the Nigerian savanna—a grassland of relatively short, feathery grasses, with scattered trees of acacia, baobab, and silk cotton. Rain totaling some 30 inches falls on Kano during five months of the year, compared with an annual rainfall of under four inches, concentrated in just one month, at

Agadez. Southward another 450 miles, and we are in the dense, humid forests that fringe the Gulf of Guinea. Here, some rain usually falls in every month of the year, but there are two especially wet seasons, and total annual rainfall is often more than 120 inches. This is the tropical-rain-forest belt, which benefits from its proximity to the ocean and from receiving water picked up over the sea by tropical maritime air masses. (Sometimes meteorological conditions allow these masses to penetrate to the Sahara, but with little benefit to the desert, because most of their water content has been lost by the time they penetrate so far inland).

The African tropical rain forest fringes the West African coast and sprawls across the Congo Basin in the heart of the continent. Similar forests occupy the Amazon Basin and the Malay-Indonesian Archipelago. These areas, which straddle the planet's equator, have the most stable of climates on land: constant warm temperatures, some rain in every month, and a high total annual rainfall. In some places the rainfall is phenomenal. In the south of the island of Fernando Po, just 1050 miles south of Agadez,

the average annual rainfall is 36 *feet* (432 inches)!

In this abundant warmth and moisture, vegetation flourishes. Instead of a scattering of small, drought-resistant succulents, giant broad-leaved trees soar to a height of 200 feet. The trees have relatively narrow trunks, often with thin, pale bark (a great contrast to the thick, fire-resistant bark of most savanna trees). Many display graceful buttresses, developed from roots at the base of the trunk, which prop up the towering plant on an often shallow root network.

Grasses are rare in the high forest, but there is a sparse growth of herbs on the dim forest floor. The vegetation is characteristically layered, so that a great block of photosynthetic tissue makes maximum use of the light falling on it. Above the herb layer, in which arums and gingers are common, is a dense understory of shrubs, and above this rises a layer of small trees (many with narrow crowns). The next story up, usually between 60 and 100 feet high, is the forest canopy, formed from the crowns of the large forest trees. Many of these crowns are in contact with one another, so that viewed from above they appear to form a rolling green sea. Sprouting up through the canopy are the occasional crowns of emergents, the forest giants 190 feet or more high.

The rain forest is evergreen. Some trees produce new leaves through most of the year, others flush during a limited season, sometimes shedding their old leaves first. The forest is never bare, and some fresh, young leaves are always present. Flowering, too, is not limited to one time of year, although there are fluctuations in the total number of flowers and fruits present at any one time. Although rain falls in every month, even the most stable equatorial climates have an uneven distribution of rainfall through the year. But the seasonal fluctuation is much less than that found in nonequatorial areas.

In temperate regions, the broad-leaved trees that are related to those of tropical forests commonly shed their leaves in the winter. This is when cold weather makes it difficult for trees to draw water from the soil, and the low light intensity provides scant energy for photosynthesis. Closer still to the polar ice cap, the forests of northern Eurasia and America are dominated by coniferous trees, with needle-shaped leaves that are drought-adapted in similar fashion to desert plants. The small surface area of needle-shaped leaves reduces water loss and their thick outer coverings protect their store of moisture through-

out the long winter, when little water can be drawn from the frosty soil. By keeping their leaves on, the conifers can at least take maximum advantage of the very short favorable growing season, for their leaves can go to work as soon as spring comes.

The tropical rain forests have no problems with frost, of course. Compared to the fraction of an ounce per square yard of annual production managed by much desert vegetation, a square yard of tropical forest has been estimated to produce up to 11 pounds net after respiration. Although these forests occupy only about four per cent of the earth's surface, they may account for 24 per cent of all its primary production. Both these figures are, however, being reduced by man's destruction of the rain forest. This high level of plant production supports a large biomass of animal life. Moreover, the tropical rain forest is rich not only in the quantity of material it contains (its standing crop), but also—and especially—in number of species, both plant and animal.

The forests of Southeast Asia are particularly rich in plants: in the Malayan forest 381 species of large tree have been counted in an area of only 57 acres; and 50 small sample plots totaling 49 acres in the forest of Borneo have been found to contain the almost incredible total of 472 tree species. It would be impossible to find one twentieth of this number in a natural northern temperate forest. Added to these trees are large numbers of shrubs and herbs and a dense population of parasitic and epiphytic plants. Epiphytic plants are plants that grow on others without parasitizing them. Orchids and ferns, for example, are common tropical forest epiphytes; they obtain a perch on the branch of a tree near the light and rely for moisture on what they can trap during rainfalls.

The animal community of the rain forest is similarly diverse. For instance, well over 1300 species of land bird breed in the forests of Venezuela and Colombia in northern South America. In a typical temperate ecosystem you would commonly find less than 150 species. Many kinds of animals that live in tropical rain forests are completely absent from other parts of the world. Yet, though the high production of the

In great contrast to the desert's paucity of life is the luxuriance of a tropical rain-forest ecosystem. This picture, taken from the banks of the Padas River on the island of Borneo, shows the key to the difference: abundance of water.

84

Left: typical of the rain forest are these South American trees, which have shallow root systems and prop up their slender, high-rising trunks with graceful buttresses. The unbranching trunks reach up through the layers of vegetation to support their spreading crowns of leaves.

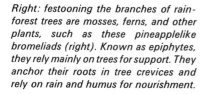

Right: festooning the branches of rainforest trees are mosses, ferns, and other plants, such as these pineapplelike bromeliads (right). Known as epiphytes, they rely mainly on trees for support. They anchor their roots in tree crevices and rely on rain and humus for nourishment.

forest supports a large number of different species, it does not support large numbers of individuals within any given species. In fact, most species are represented by a rather small number of individuals. This applies to the plants and trees as much as to the animals that live in them.

Why should this be? Why are so many different species found in tropical rain forests? A common answer is that the enormous diversity is a result of the stability of the climate. In Chapter 4 we mentioned the concept of the ecological niche— the special "job" that a species performs in an ecosystem. In the tropical rain forest, climatic stability can make a very great number of niches available. A plant or animal can become adapted to a special way of life under particular conditions that remain relatively unchanged, whereas such specialization is usually not possible in more seasonal climates. There is no "job" for a permanently active insect-eater in the trees of a northern deciduous forest. For half the year the trees are leafless and nearly without insect life; and so the insect-eater must migrate to a different environment during this unfavorable period, or must either hibernate or switch to other dietary items at other levels of the forest. But in the tropical forest such a creature can specialize in eating particular types and sizes of insects living in particular parts of the eco-

system all through the year. A closely related insect-eater can meanwhile specialize in feeding on a different type of insect, or in a different part of the habitat. Each of these two species can therefore fill its own niche without interruption or undue competition.

Although the stable climate obviously permits many similar species to exist without directly competing, it may not be the full explanation of how and why the diversity arose. Most moist tropical forests seem to have been affected by the dramatic climatic changes of the Pleistocene epoch (which included the Northern Ice Ages), as well as by very long-term patterns of change in the positions of the continents and in worldwide climate. Changing rainfall distribution and fluctuations in sea level during the Pleistocene epoch probably brought about a good deal of fragmentation and subsequent rejoining of tropical forests, so playing an important part in producing new species. The animal populations of the fragmented forests would have been broken up into several isolated groups, which probably evolved into distinct species—distinct enough, at least, to coexist when forest expansion joined them together again.

The tropical forest community is certainly diverse, whatever the reasons for its diversity may be. For instance, high above the ground in the forest canopy, where most of the ecosystem's food supply is produced throughout the year, lives a tremendous range of specialized arboreal animals, many of which never even visit the forest floor. We can see an example of the variety of tropical forest animals and their niches in an examination of the monkeys of the African rain forest. Although some monkeys live in grassland and scrub, the great majority of species are forest-dwellers—efficient climbers with grasping hands and feet, good vision, and long, balancing tails (which some South and Central American species use as a fifth grasping limb). The African forest is not identical across its several thousand miles of continental spread, and so there are rarely more than half a dozen monkey species in any one area. The forest as a whole, however, contains about 30 distinct monkey species.

In parts of the Kibale Forest of western Uganda, on the eastern fringes of the main Congo forest block, six monkey species coexist: the redtail monkey, the blue monkey, l'Hoest's monkey, the gray-cheeked mangabey, the red colobus, and the black-and-white colobus. The six species are all members of the same zoological family, are quite similar in size, and inhabit the same part of the forest; but they occupy slightly different niches, mainly because their diets differ. Thus the six species do not constantly compete with one another. Indeed a single tree may at one time contain a peaceful association of two or three types of monkey in any combination.

The redtail monkeys, with white nose spots and rusty-red tails, are the smallest of the six. Their diet includes great quantities of insects and insect larvae, together with some fruits and flowers and a small number of leaves. They move about rapidly, within a limited part of the forest, in groups of about 20, often feeding quite low down and foraging for insects as they go. The range of the group—the area they live in permanently—covers about 50 acres of forest. In contrast, a group of blue monkeys, which is of similar size, has a range of at least 200 acres. The blues are longer-haired than the redtails and actually steely-gray rather than blue in color. They tend to feed higher in the forest canopy than the redtails and although their diet includes insects, they do not eat nearly as many as the redtails. Instead, the blues feed largely on fruit. This may explain their need to move about over such a large area. Trees in the forest bear fruit

only occasionally, and so the blue monkeys have to search rather widely for their supply.

Mangabeys also eat large amounts of fruit and they too, with a group size of only about 15, have a very large range. Theirs, in fact, is enormous: nearly 1000 acres. Bark provides another significant part of the mangabey diet, as do small animals; the mangabeys occasionally eat snakes and other small vertebrates as well as insects. They generally feed in the lower part of the forest canopy.

The redtails, blues, and mangabeys all move through the forest by walking along branches and jumping between trees, whereas the two species of colobus move along the branches in a bounding style and make great, crashing leaps from tree to tree. Both species of colobus—the red and black-and-white—are about the same size as the mangabeys (the male weighs about 22 pounds). Red colobus monkeys move within a range of at least 90 acres, and in much larger groups than those of the other forest monkeys. Normal groups of red colobus in the Kibale Forest contain about 50 monkeys—a very high population density. They eat some insects and fruit, but their main items of food are the buds of leaves and flowers, and the stems of leaves. Their diet is in fact very diverse (they feed on dozens of tree species), and they find much of it in the canopy and in emergent trees.

The black-and-whites have a much less varied diet than the reds: they concentrate mainly on the leaves of a small number of trees from the middle story, where they can find and devour great quantities of the blades of large leaves, together with some buds, fruits, and flowers. The only monkeys in the Kibale Forest that eat no animal material at all, these black-and-whites live in small groups of about 10 animals, and they range over a very limited area, generally no more than 40 acres. Again, the extent of range is probably related to diet, because they do not have to travel great distances in search of a fruit crop, as do the mangabeys and blues.

The separation of these five monkey species is not absolute. There are times when they will all eat fruits of the same tree species in the same place. However, their combined diets cover scores of plant and animal species and many different plant parts. By emphasizing different sections of this varied menu in their individual diets—and in a habitat where food is abundant— the five species are able to coexist.

The head of a flower mantis, an inhabitant of tropical mountain rain forests. The mantis's flowerlike appearance may attract nectar-sucking insects, which it seizes and eats.

An Indian moon moth of the Himalayas. Like the other small animals of the tropical forests on these pages, each of the horde of moth and butterfly species occupies a special niche.

Tarsiers, related distantly to the ancestors of monkeys and apes, live at low levels in the Southeast Asian forests, searching at night for their insect food on the trunks of trees.

A gigantic hairy spider of the South American rain forest. These hunting spiders are often called "bird-eating" spiders but they probably feed mostly on lizards and insects.

Variety in the Rain Forest

In the Kibale Forest of Uganda, six species of monkey live peacefully together. They concentrate on different foods and live in groups of different sizes. Groups of some species range over a large area, while others are much more restricted in their movements. The species tend to feed at different heights in the forest, one species, the l'Hoest's monkey, spending almost all its time on the ground.

Red colobus
Colobus badius

Group Size : 50
Range Size : 90 acres
Diet : mostly young leaves, buds of flowers and leaves, stalks of mature leaves; also some insects, fruits, and flowers.

Blue monkey *Cercopithecus mitis*

Group Size : 20
Range Size : 200 acres
Diet : fruits, small insects, flowers, and flower buds; also a few young leaves.

Redtail
Cercopithecus ascanius

Group Size : 20
Range Size : 50 acres
Diet : small insects and fruit; also some flowers and their buds.

Gray-cheeked mangabey

Cercocebus albigena

The illustration shows, in diagrammatic form, the relationship between the areas of forest used by groups of five of the six monkeys. The largest area, used by the mangabey, is equivalent to 1000 acres and the smallest area, used by the black-and-white colobus, is equivalent to 40 acres.

Group Size: 15
Range Size: 1000 acres
Diet: inner bark of trees, fruits, insects, and other small animals; also a few leaves and flowers.

Black-and-white colobus

Colobus guereza

Group Size: 10
Range Size: 40 acres
Diet: young leaves, mature leaves; also some buds, flowers, and fruits.

L'Hoest's monkey *Cercopithecus lhoesti*

Group Size: 20
Range Size: unknown
Diet: fruits and shoots of herbs, mushrooms, insects.

The sixth type of Kibale Forest monkey, l'Hoest's monkey (closely related to the redtail and blue monkeys), is unlike the other species, in that it spends most of its time on the forest floor or in low vegetation, moving stealthily in groups of about 20 and eating insects, fruits, mushrooms and the shoots of herbs. Living in a different forest layer, l'Hoest's monkey competes hardly at all with the five other species.

The monkeys are but one type of consumer in the African rain forest. They act at several levels in the food chain, eating producers, primary consumers, and a few secondary consumers. They must compete for their food at each level with a huge variety of other species, notably with birds (such as hornbills and turacos) for fruit, with caterpillars for leaves, and with insectivorous birds and mammals for insects. And they are themselves preyed on by eagles. They are just one example of the great variety of life forms that have come to exist in a highly productive ecosystem, flourishing under the most favorable of conditions on land.

This ecosystem contrasts strikingly with that of the desert, where the climate, which is extremely unfavorable for life, has given rise to a community very different in density and appearance. We turn now to still another habitat: water, the original home of life on our planet. And again we shall see that although an aquatic ecosystem presents living things with problems and opportunities quite unlike those faced on land, it nevertheless has the same basic attributes as the desert, the tropical rain forest, and every other major land ecosystem.

Earth's waters are by no means uniform, however. They present as wide a range of distinct environments for life as does the land. The oceans, whose waters are contiguous, can be regarded as one very large-scale ecosystem, but it is more convenient to examine the oceans in smaller sections. We can do this by classifying them according to geographical units or environmental divisions, such as seashores (which can themselves be subdivided according to their location and the amount of sand, mud, and rock they bear),

the seabed (which is very different close inshore as compared with the deep ocean trenches), the open ocean (divisible into inshore, mid-ocean, polar, temperate, and tropical waters), and so on. As for the inland waters, they can be either fresh or salty, and their variety embraces such environments as lakes, ponds, swamps, streams, and rivers.

A freshwater river is a very different ecosystem from any section of open ocean. About the only physical feature the river and ocean have in common—unless they are geographically close—is an abundance of water. The water that rains onto the land comes originally from the oceans, carried through the atmosphere by winds. After falling onto the land, most of it finds its way back to the ocean, draining first through the soil, then bubbling down in streams and finally flowing sedately out of wide river mouths into its former home. During this course it may be held for months or years in lakes. The water erodes the

Above: a wandering albatross spreads its wings over the Southern Ocean. The slowly drifting water-mass carries with it vast numbers of tiny floating plants and animals, the plankton, which is the food supply for the fish eaten by the albatross.

Right: this fast-flowing Alpine river presents living organisms with problems very different from those of the open ocean. In its cold, violent upper reaches few plants can gain a foothold. Thus most of the food for the small animals sheltering among the rocks comes from the leaves and other matter from the land.

In the middle reaches of rivers, where the current is moderate, the water clear, and the bottom not too deep, rooted plants, such as these Sparganium *(lower)* and Potamogeton *(center)*, can take hold and flourish despite the drag of water.

land as it travels across it, carrying particles and dissolved mineral salts into the sea. As a consequence of the leaching of salts from the land, the sea has a high concentration of dissolved salts, so that even the common physical feature of river and ocean—their water—is not *truly* common. The water of the ocean cannot, then, be a closely similar environment for life to that of the river.

The oceans are probably not getting increasingly salty, however. Many salts are blown back to the land by wind, and others are deposited in ocean sediments and carried back above the high-water mark by geological processes. After an increase in salinity in the early part of earth's history, the oceans have probably been in a fairly stable state for at least 500 million years. Different parts of the ocean, though, like different rivers, contain different amounts of salts. For instance, the Baltic Sea contains only about 7 parts of dissolved salts per 1000 parts of water, compared with more than 40 parts per 1000 in the Red Sea and an overall average of about 35 parts. Most rivers contain less than one part per 1000, and many have less than that—often less than one tenth of a part.

Many of the plants and animals living in the sea have evolved there, and their tissues generally contain a concentration of salts that nearly matches the salt content of the surrounding water. The distant ancestors of most freshwater organisms invaded the freshwater habitat from the sea or the land (the land-dwellers themselves having descended from early aquatic forms), and their tissues inevitably contain a greater concentration of salts than their dilute environment. Freshwater animals have to employ pumping mechanisms—excretory organs (such as the kidneys of fish) or tiny vacuoles (such as those of protozoans)—to rid their bodies of the large amount of water that flows into their cells by *osmosis* (the movement of a solvent into a more concentrated solution through a semipermeable membrane). Plants, on the other hand, do not have pumping mechanisms, but their cells do have rigid walls, which both stop them from swelling and exert a back pressure that prevents water influx.

However, some marine animals, such as bony fish, evolved in fresh water from originally marine ancestors. The tissues of a bony fish contain, as a result, a smaller concentration of salts than the ocean does, and so the fish tends to lose

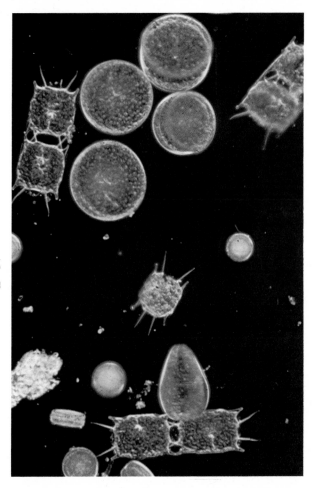

These single-celled plants are diatoms, typical members of the plant plankton of the open ocean. Diatoms have a silica shell divided into two parts, forming a protective box and lid.

water to its surroundings. To counteract this loss, it must drink the sea water and get rid of excess salts in its feces and by excretion through the gills. Only a few animals can adapt readily to life in both fresh and salt water. Among these are the several species of salmon that breed in rivers but feed mostly at sea. For part of their life they must bale water out through their kidneys; for the other part, they drink sea water and excrete the excess salt from their gills.

Apart from the amount of salt concentration, there is one other obvious major difference between river ecosystems and those of the open ocean: whereas the water of open ocean drifts slowly in currents and has an average depth of 12,500 feet, rivers flow, often fast, and are rarely very deep. This is, of course, why free-floating plants are rare in rivers; the plants that do grow in running water are in constant danger of being swept away. In the upper reaches, where the water may flow swiftly down steep slopes, the current is generally too strong for most plants to gain foothold other than at the edge of the stream or on rock surfaces above the water. (Some algae do, however, manage to spread a slimy green film over the larger stones.) Lower in the river's course, where its current has slackened and the bottom sediment is thicker and less disturbed, rich growths of rooted plants may take hold. In shallow rivers, such plants— water crowfoot, pondweeds, and eelgrass, for example—may be rooted in the center of the waterway, managing to reach up from the bottom into well-lighted water; but in large rivers you can generally find rooted vegetation—particularly reeds—only at the fringes.

The open ocean presents producers with very different problems. The bottom in which plants might root is often many thousands of feet away from the zone at the surface through which sufficient light penetrates for photosynthesis (for this zone is very thin, usually less than 350 feet thick). To survive in this ecosystem, photosynthetic plants must float, drifting slowly with the ocean currents. The ideal plant form for such an existence is, as we saw in Chapter 2, the single cell. As the available resources under water are evenly spread, there is little need of any specialized support, water-gathering, and food-producing organs. The smaller the plant's volume in relation to its surface area, the slower it will sink (and to sink from the lighted zone is to die) and the more easily will it exchange materials with its environment. The great majority of open-ocean floating plants, the phytoplankton, are single-celled algae: diatoms, dinoflagellates, and coccolithophores. To slow the rate at which they sink, many such plants have a surface area increased by outgrowths, whereas others contain oil droplets for buoyancy.

The aquatic ecosystems of neither the rivers nor open oceans are as efficient at fixing solar energy as are many ecosystems on land. Water transmits light poorly: the water surface reflects much of the light, and the water absorbs much that is not reflected. Even in the clearest seas, less than one per cent of light falling on the surface reaches a depth of some 350 feet. Naturally, the restricted amount of solar energy limits plant growth, and plants may also be denied some of the other materials essential for their survival. For instance, although carbon dioxide is usually abundant, mineral salts are not always present in sufficient concentrations.

In rivers, the mineral content is related to the soil and rocks through and over which the river drains. Thus, if any mineral is scarce in the substrate, the river is likely to have a poor supply of that mineral, and this will affect its living community. The concentration of minerals may also be weakened by a heavy rainfall, which makes the river more dilute. But the mineral supply situation in the ocean can be much more precarious for plant life. True, the sea contains a greater overall concentration of mineral salts than does fresh water, but rivers are constantly being supplied, whereas the ocean lacks adequate quantities of certain minerals. Some of the minerals that are essential for plant growth, such as phosphorus and nitrates, are nowhere abundant —and may be in their lowest concentration in the very surface water where the plants live.

The food chain moves downward: organic detritus and the minerals it contains accumulate on the bottom. Mineral salts can diffuse through water only slowly, so that at the surface of the open sea—far from the bottom and far from rivers emptying their cargos at the coast—phosphorus and nitrates are usually scarce. As a result, only a thin phytoplankton population can be supported. This is especially significant in warm seas, where the water can become layered, with little exchange between the warm water floating at the surface and the heavier, mineral-rich colder water below.

In some warm oceans, on the other hand (the part of the Pacific off the coast of Peru is an example), upwelling currents bring high concentrations of minerals to the surface: and in such waters the phytoplankton flourishes. This high productivity supports a dense population of marine animals, including the fish and seabirds that form the basis of the anchovy and guano industries. Net primary production may reach about 12 ounces a year per square yard in such a situation, as compared with less than two ounces per square yard in most open-ocean ecosystems.

Rivers vary greatly in primary productivity.

Land and water meet in central Africa. Rivers and lakes get much of their food supply from the land. In Africa hippopotamuses graze on land at night and then return at daylight to the water, where their feces nourish many other aquatic organisms.

Left: freshwater shrimps are abundant animals in the fast-flowing temperate streams. They feed largely on detritus.

In the open ocean the chief primary consumers are small floating animals, mostly crustaceans (below), which graze on the even smaller floating plants.

Much depends on the actual section of the course of a given river and the climatic and soil conditions to which it is subjected. As an extreme example of high productivity, consider one remarkable stream in Florida: most of the primary production is carried out by eelgrass coated with algae, and this amounts to an estimated 12 pounds a year per square yard, of which about 40 per cent (five pounds) is the net production available to consumers. Obviously, this is an exceptionally high production, similar to the average for tropical rain forest. But it occurs in a particularly favorable situation with clear, shallow water, a moderate flow, high mineral concentration, and constant temperatures.

In many rivers, much of the energy supply, especially in the upper reaches, comes not from the production of plants that grow in the water, but from material that either falls into the river or is washed from its banks. In the headwaters of a stream, where there is a minimal amount of rooted vegetation, a large amount of the food supply for the system is debris of this sort. Even in the lowland reaches, a significant proportion of the energy entering a river is material washed down from higher up or brought in from the surrounding terrestrial ecosystem. Hippopotamuses in African rivers provide us with a special example of this process. At night, the hippos emerge from their watery homes to graze on land, returning to the water as daylight comes. Much of their defecation occurs in the water, so that the aquatic system—often a lowland river—is enriched with material produced on the land.

Leaf-fall is another example of terrestrial production that feeds a river ecosystem. Many consumers in rivers are, therefore, detritus feeders. To exploit efficiently the flow of detritus, as well as the bacteria and fungi decomposing it, they must be able to resist disturbance by the current. In fast-flowing headwaters in Europe and America, the chief consumers of detritus are the flattened larvae of stoneflies and mayflies, which cling to the surface of stones, and the flattened freshwater shrimp *Gammarus*. Also in these streams lives the larva of a caddisfly, *Hydropsyche*, which spins a silk net on the underside of a stone. The larva lurks inside the net, which has a wide opening leading to a closed tunnel, and consumes the detritus and small animals swept in by the current and trapped. Lower down the river, where the current is sluggish, rooted plants are more abundant and a softer bottom is available for burrowing. There the animal population increases dramatically.

Unlike rivers, the open ocean gets little of its energy from terrestrial debris. But the upper layers of the sea supply food to the dark depths much as the land feeds unproductive rivers. It can, indeed, be said that an ocean-bottom community exists on the detritus raining from above. Bacteria and filter feeders consume the small particles—dead plankton and the remains of predators' meals from high above—and large fish patrol wide areas in search of larger corpses. When one of these falls to the bottom, the big fish move in like vultures.

As has already been indicated, rivers and oceans share the influence of gravity in their food chains. In other words, the food chain moves downward as each successive link tends to rely partly on food that falls or swims from higher up. In the sea, vertical migrations between different depths are common, and such migrations help to support quite a large population of predators in the twilight zone beneath the layer of plant production. Many crustaceans, which are the chief browsers on phytoplankton, feed at night in surface waters, but descend to depths of some 650 feet or more by day. One reason for this behavior may be that it permits them to take advantage of the differences in current speed between the surface and the water below. By switching in and out of the faster-moving surface water, the animals constantly change their position in relation to the rest of the plankton and can browse on a different plant community every night. These migrations, then, bring food materials down from their point of production to within reach of the upward movements of deeper-living consumers.

In the permanent dark of the deep sea (the bathypelagic zone) lives a sparse population of chiefly solitary predatory fish, which feed mostly on other predatory fish. Competition here is intense. Some of the fish have luminous lures with which to entice their prey and many have huge mouths and highly distensible stomachs that allow them to make the most of each of their usually infrequent meals.

The living things least affected by the great differences between the river and open-ocean ecosystems are the top carnivores, the most adaptable and opportunistic of species in all environments. Certainly, it is only in the ocean depths that we find such bizarre fish as those just described. And there are giant squids,

Right: the pike lurks among reeds and other thick vegetation in fresh waters, dashing out to seize unwary animals in its elongated jaws. The jaws carry backward-curving teeth that stop the prey escaping before it is swallowed.

Below: Goliath herons dispute over a fish. The long, stabbing bills of these huge birds (five feet high) of African inland waters resemble other predatory water birds. Although they display a variety of shapes and feeding techniques, many (including the Goliath) hunt by standing in the water, striking at fish and frogs that venture too close.

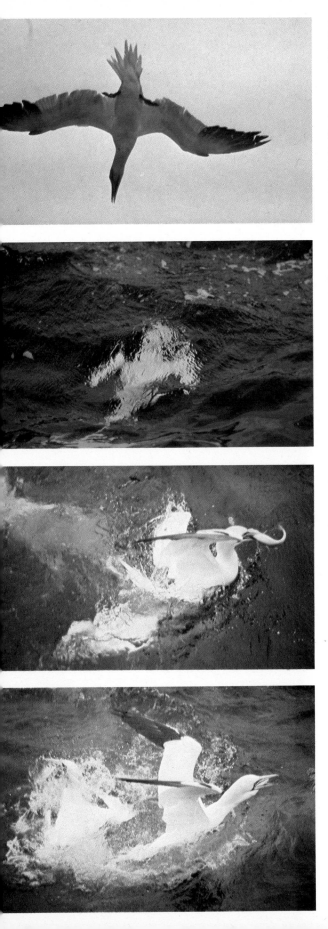

Left (top to bottom): the northern gannet will drop from over 50 feet to snatch fish from the surface of the sea, folding its wings as it dives. Predatory birds play a significant role near the top of food chains in both sea and fresh water.

whales, and seals in the sea that are almost entirely absent from rivers. However, all these animals are large, many are fast-moving, and all but the squids are vertebrates. In rivers, too, the top carnivores tend to be big, swift, and vertebrate: fast-swimming fish, and active birds, reptiles, and mammals that, like many of their relatives in the sea, retain some attachment to the land. In fact, individuals of some top carnivore species, such as the fish-eating cormorant, may feed at different times in both inland waters and the ocean.

Desert, rain forest, open sea, river—these four very dissimilar sorts of ecosystem are distinguished from one another primarily because of differences in the amount and nature of the water each contains or receives. All four can exist at the same latitude: all four may get equal amounts of light energy through the year: yet they remain completely different. But physical factors other than water can have great effects on the nature of an ecological unit. Because temperature and light change with latitude, there are vast ecosystem variations between the equator and the poles. Similar variations, resulting mostly from declining temperature, occur between the base and summit of a tropical mountain. Vegetation on land is strongly influenced by the soil, and marine ecosystems are greatly affected by depth and by their proximity to coasts. But despite the differences, all ecosystems have much in common.

Every ecosystem is an inextricable web of life, inanimate matter, and weather. Their contents are diverse, but they have a unity in that they all belong to the whole earth ecosystem—the biosphere. As a functional ecological unit, every ecosystem is relatively balanced, but the balance is of a dynamic nature, with individual plants and animals constantly interacting, reproducing, and dying, and with physical conditions constantly changing. It is within such arenas—seemingly permanent, yet always in flux—that the processes of evolution take place.

Ecosystems and Time

The movement of continents, the alteration of weather patterns, and the evolution of plants and animals, these are some of the factors that have contributed to the changing of ecosystems with the passage of time. The elements in the balance of an ecosystem do not remain the same. Given enough time, all the elements alter. Three hundred million years ago in the Carboniferous era, giant horsetails, club mosses, and ferns 50 to 100 feet high flourished in a warm, moist environment on land that is now temperate Europe and North America. In this swampland dwelt massive fish-eating amphibians 15 feet long and dragonflies with 30-inch wingspans. The swamps became fossilized as coal; and today the ecosystem above the fossil coal is mostly temperate deciduous forest, greatly modified by human cultivation and urbanization—urbanization that has been established partly as a direct result of the mining of the coal.

Ecosystems differ in their stability. Until agricultural man began to modify the natural world dramatically, an ecosystem such as mature, temperate deciduous forest remained little changed in its gross structure for many hundreds of years, a period much longer than the life-span of most of its inhabitants. Forests that are developing to this mature state following some disruption change rapidly, however; they differ from year to year in their structure and appearance. At the other extreme are some of the forests of the North American Pacific coast. Bristlecone pines, redwoods, and giant sequoias growing there today are thousands of years old, and are living evidence of an extremely stable system. For their part, the seas have always been considerably more stable than any land area. The basic conditions of life, and types of animal over large areas of the seabed, have remained relatively unchanged for millions of years.

Even the most stable ecosystem is subjected

Seams of coal—the fossilized remains of swamp forest vegetation—exposed in a rock face in the desolate, frozen Antarctic testify to great ecosystem changes. Since this coal was laid down, about 250 million years ago, the Antarctic continent has probably drifted south from more central, warmer latitudes.

to some slight degree of short-term change because of the seasonal variations of climate and the alternation of day and night that occur as the spinning earth orbits the sun. These changes can be seen and felt in all the lighted areas of the earth exposed to the shifting air masses of the atmosphere, but not in the depths of the sea, which remain constantly dark with little temperature change. But the deep waters rely for their energy supply on happenings in the lighted zones above, and so any changes at the surface must inevitably have some effect on the depths. Another short-term change of significance in the oceans is the tidal rhythm, set up by the interaction of the gravitational pulls of sun and moon on the surface waters. The tides have their most marked effect on the plants and animals that live on the seashore below the high-water mark, which must adapt to a cycle of alternate immersion in water and exposure to air as well as to the cycle of day and night.

The regular sequence of day and night is the most universal of the environmental changes induced by time. One always follows the other, although the actual length of each varies with latitude and season. Close to the equator, day and night are each about 12 hours long throughout the year, but inside the Arctic and Antarctic circles the "day" is about six months long, and several weeks are completely dark. A few plants and animals live permanently in these extreme polar conditions, whereas others spend just part of the long "day" there. But the majority of living things pass all their lives in ecosystems where several hours of daylight are followed by several hours of darkness.

Although an ecosystem as a whole does not change very much from one day to the next, its environmental conditions and the activities of its organisms can change markedly between day and night. Daytime sunlight allows photosynthesis to occur in plants and raises the temperature of the environment and the rate of water evaporation from it. We have already seen the great changes that occur during a 24-hour period in the desert, where many small animals are active only at night and plant photosynthesis slows in the heat of the day as stomata close. In the open ocean, the phytoplankton produces photosynthetic products by day, but many of these products do not reach the next link in the food chain until dusk or at night, when planktonic

Below: bongo antelopes photographed by night at a water hole. Most large African antelopes are grassland-dwellers that move and feed mainly in the cool of the early morning and late afternoon, but the forest-living bongos are more active at night.

crustaceans migrate to surface waters from below. Food chains clearly work on a time basis, because plants can carry out photosynthesis only during the day, whereas many consumers are most active at night.

We human beings are day-active, or diurnal, animals and most of us are not fully aware of the active night-live of many nocturnal animals. Nighttime activity helps to conserve the moisture in their bodies. Unlike plants, which are rooted in supplies of water, terrestrial animals must carry their water around in their bodies. Especially in the middle of the day, they can lose a great deal of water through evaporation and this is particularly true of such animals as birds and mammals, which keep constant body temperatures and rely partly on the evaporation of water from their surfaces to prevent overheating. Because, then, of the need to conserve moisture, few animals are active at midday in any tropical land ecosystem. Among the exceptions are African antelopes such as the impala, which may drink and whose females usually give birth in the heat of the day. There is good reason for this: when drinking or giving birth, the antelopes are vulnerable to predators; but the danger is less at midday when predators themselves are likely to be resting.

Tropical rain forest exhibits some of the clearest day-to-night variations in its ecology. Specialization for a particular niche is affected by such variations. Obviously, if an animal is highly specialized for a particular daytime role, it is going to be less efficient at doing the same thing by night. So when dusk falls, the animal community in the rain forest undergoes a massive switch-over; the night shift is composed of many species that are inactive in the daytime. For example, because the eyes of African forest monkeys are, like ours, adapted to daylight, they see poorly at night, and it would be dangerous for them to move about through the trees. When night falls and they go to sleep, other mammals become active. Fruit bats leave their roosting trees to devour figs and other fruits (which the monkeys were eating by daylight), and scaly-tailed flying squirrels emerge from their holes to browse on leaves and flowers. Bush babies leap among the trees, grabbing insects that are feeding nocturnally on the vegetation, and insect-eating bats swoop on other insects in flight.

The creatures that lead an active night life are, of course, physically equipped for it. Bush babies have exceptionally large eyes, with a reflective layer behind the retina. Any light that enters the eye stimulates the sensitive retinal cells twice, intensifying the images that the animal picks up in the gloomy forest. And the insect-eating bats have a sonar system that enables them to feed without using their eyes at all.

Many of the insects that the nocturnal animals consume are quite different from the types eaten by such diurnal insectivores as birds. Moths replace butterflies, and mosquitos replace biting tabanid flies. Higher up the food chain, predatory hawks and eagles give way to large-eyed owls; on the forest floor civets take over from mongooses. The human hunter sleeps, and the

Broad-leaved trees growing in parts of the temperate zone shed their leaves before they are damaged by winter frosts. In winter, when much of their food is lost, many deciduous-forest animals become inactive or migrate, but soil life is still active, nourished by the thick leaf litter. With the longer, warmer days of spring, leaf buds open on the trees and fresh green vegetation sprouts from the forest floor, so increasing enormously the food available for both awakening and returning animals.

leopard hunts. Through the night, lowland tropical forest remains warm and becomes even more humid than in daylight. This is the most favorable time for small animals without waterproof skins, such as tree frogs, to move about without threat of desiccation.

In the tropical rain forest, the difference in the functioning parts of the ecosystem is greater between day and night than it is at the same time of day in different seasons. In the temperate forests to the north, however, seasonal differences—which occur as the earth is first tilted away from the sun and then toward it—are much more significant. In midsummer a deciduous broad-leaved woodland has much in common with a tropical forest. It frequently contains several layers of vegetation, and the animal population is dense and varied, for many species have produced offspring and many species have migrated from distant lands. Caterpillars browse on the foliage of trees, and squirrels and other rodents eat the

Many northern animals hibernate through the cold winter months when food is in short supply. Among the hibernators are those that include a large number of insects in their diet, such as the garden dormice (below) and the greater horseshoe bats (right), because insects are inactive in the cold.

fruit; deer browse in the understory, and small mammals eat vegetation on the woodland floor. A variety of birds, especially warblers, feeds on the insect population in the trees, and shrews and hedgehogs eat insects on the ground. Hawks, owls, and carnivorous mammals prey on the smaller birds and mammals. Most of the mammals have broad diets, for the availability of different types of food is constantly changing. Bears eat a great variety of fruit, roots, honey, insects, small and large animals, and carrion.

With the approach of winter, the deciduous forest changes completely. As a protection against the coming cold, the broad-leaved trees

and shrubs shed their leaves. This helps to prevent frost damage and a water loss that cannot be easily replaced from the soil when temperatures are low. Many small herbs die away completely in the autumn; they are the annuals, which, like the desert ephemerals, survive the winter as seeds. Other plants die above ground but remain alive underground with a supply of food reserves —within a tuber or bulb, for instance. Even the plants that retain green leaves above the surface produce little food during the short, dim, cold winter days. Thus the animal population that flourished in midsummer is faced with a huge drop in available food.

Some of the primary consumers of the deciduous forest solve the food problem by becoming dormant. Most insects pass the winter in a state of suspended animation (which zoologists call the *diapause*). They may do this as pupae in the soil, as eggs, or as adults. Many small animals and insects die during the winter. Some mammal species, such as dormice and ground squirrels, go into hibernation in a nest or burrow. The hibernating animal becomes torpid, its heart and breathing rates fall, and its temperature drops to a level close to that of the environment. If the temperature drops dangerously low, the creature automatically warms up, sometimes awakening

109

in the process. In this way the hibernator consumes little energy—but it does use some, derived from extra layers of fat laid down during the bountiful summer. (Rodents are not the only hibernators. Bats and bears of the temperate forest are other mammals that sleep through the winter.)

Some consumers do not give up all their activities, though. For instance, tree squirrels do not hibernate, but simply become less active. They rely on fruits stored before the winter and on what food is still available on the ground. There remains plenty of food *in* the ground, too, and many consumers keep active by relying on the soil for food. The earth has been enriched by the recent leaf fall. It also contains pupating insects and plant storage organs, and because temperature conditions are more stable in the soil than in the air, the soil's community usually continues to function.

The disappearance of many primary consumers obviously affects secondary consumers. Especially hard-hit are those that feed mainly on flying insects, which are almost totally absent from the winter forest. Insectivorous bats hibernate; some of them travel hundreds of miles to overwinter in ancestral caves. Such directed long-distance movements, which are common seasonal phenomena in the animal world, are known as *migrations*. The most notable of the migrants are the insectivorous birds of the northern forests. With the approach of winter, multitudes of birds from the forests of the north start to wing their way southward. For example, those from northern Europe and Asia travel down to the grasslands of Africa, many including in their journeys a crossing high above the Sahara, in 30–40 hours of nonstop flight. Not all have the same route, however: some keep to the coast, some hop from oasis to oasis, some follow the Nile. One estimate is that

Snow geese spend the summer on the tundra of the American Arctic, exploiting the ecosystem's brief outburst of productiveness. As the days shorten and their food supply dwindles, the geese migrate south to winter on warm southern coasts.

Right: not only arctic animals migrate. As the dry season comes to the East African Serengeti area, wildebeest migrate in a mass from the open short grasslands they have been grazing into country that is lower-lying, more wooded, and better watered.

some 5000 million land birds (together with huge numbers of water birds) set off from Europe and northern Asia to sub-Saharan Africa each fall. Predominant among these migrants are the warblers; other groups include nightingales, swallows, shrikes, flycatchers, and cuckoos.

The migrants often find themselves in harsh drought conditions in Africa, where they must compete with a large population of resident birds. (It must be remembered that migrations are not just an element of the seasonal ecosystem change in the cold lands; obviously, the migrants affect also the systems to which they migrate.)

The migrants fly south not only from the deciduous forests, but also from the even harsher and colder ecosystems of coniferous forest and tundra farther north. These areas undergo even more extreme annual variation than the temperate forest. The tundra, which fringes the Arctic Ocean in the north of Eurasia and America and extends several hundred miles south into the continents, is a cold desert. Water is scarce not because there is no precipitation, but because of the cold, which freezes liquid in the soil and causes more snow than rain to fall. Through the long, dark winter, the ground is usually completely covered with snow, the rivers and lakes with ice. Lakes may be iced over from September through June or July. Much of the soil water remains permanently frozen: for example, the permafrost at Nelson River in Manitoba (Canada) is 23 feet thick. Only July and August may be completely frost-free—and then only the top layer of soil thaws. No trees can grow on the high tundra; the vegetation is dominated by lichens, grasses, and such heath plants as heather, bilberry, and crowberry. In southern areas there may be a thick scrub of dwarf birches and willows.

Almost all birds of the tundra are migratory.

Many species migrate southward in winter to temperate latitudes, and many others go farther: from the Eurasian tundra to Africa, and from the American tundra to Central and South America. The American golden plover travels 6000 miles from the tundra, passing across Labrador and then over the Atlantic, ending up in southern Brazil and Uruguay. It returns to the Arctic in the following year by an overland route. Wheatears from the tundra of Greenland cross as much as 1500 miles of the Atlantic, then fly southward through western Europe, across the Mediterranean and the western Sahara, to winter feeding grounds in the African savannas. The mean temperature of the savannas in winter is 74°F as compared with 36°F in their summer home. Arctic terns, some of which nest in the northern tundra, travel south in winter to the seas of the Antarctic and fly back north in spring. This round trip has been estimated to involve flying for 24 hours a day for eight months of the year.

The tundra is the summer home, too, of migratory mammals. North American caribou may travel over 1000 miles a year. Those that spend the summer on the Canadian tundra known as the Barren Grounds go south to the margins of the coniferous forest in winter. During these migrations, in which hundreds of thousands of deer used to participate (in the days when caribou were more plentiful than they now are), the deer will swim across wide rivers and lakes. Reindeer (belonging to the same species as caribou) perform similar migrations in Siberia and in those parts of Europe where they still run wild. Like many other tundra animals, caribou and reindeer in the northern part of their range grow near-white coats in the winter time, thus camouflaging themselves from wolves.

Lemmings pass the Arctic winter under the snow, sometimes after migrating a short distance from their summer home on the tundra. Such seasonal movements are different from the mass migrations for which lemmings are famous. These irruptions occur when population densities are very high and the lemmings in search of summer homes find all space already occupied.

The fluctuations in numbers that are so typical of lemming populations provide an example of another aspect of the effect of time on ecosystems. Unhampered reproduction could in theory lead to a situation in which the population of any living thing would grow far beyond the available resources of its environment. Over

wide areas of the planet, man is increasing rapidly in numbers and destroying resources that are not being replaced, but in natural ecosystems, the numbers of an animal species do not increase indefinitely; they remain fairly constant, or fluctuate around some average level. The probable ways in which population numbers are regulated have been argued over by ecologists for many years.

Clearly, if an animal population increases to a level where it exhausts all available resources — food, for instance — it will no longer reproduce successfully, and its numbers will fall. Most wild populations, however, do not grow to a point where they exhaust all available resources. Several factors intervene to prevent this from happening. One such controlling factor is the weather, which especially affects insects in harsh environments. Where severe droughts or cold spells inevitably occur every few years, all but a few members of many populations are killed off. Only those finding a favorable refuge are able to survive under such conditions.

Competition among different species can also be a potent factor in population control. One result of competition, as has been pointed out, is that different species occupy separate niches; thus, even in a richly provident environment only a small portion of the area's food is consumed by any one species. Before this supply is completely used up it may well become so thinly spread that the consumer is forced to use up more energy in gathering food than it gets from the food. Resultant death or failure to reproduce successfully then limits the population size of that species. The number of animals that fail to reproduce will depend on the total size of the population, for the larger the population, the less available food per individual. This can be considered as density-dependent population regulation (as compared to the density-independent effects of the weather). Density-dependent regulation can come into play, too, if individuals cannot find adequate breeding sites, shelters, or other necessities because of high population density. Territorial behavior can also be seen as exerting a density-dependent type of control on numbers.

Predators cut down the population size of their

Alaskan caribou migrating southward in late July from the high tundra through terrain criss-crossed by streams of glacial meltwater. Their winter diet contains large quantities of lichens.

prey, though this is probably of more significance for small animals than for vertebrates. Mammal predators commonly hunt and kill sick or weak individuals of their prey, which would very likely die anyway. But they may also eat many young animals and so have a considerable effect on the size of a prey population even if they do not ultimately control its numbers.

The control of size of a particular animal population can, in fact, involve the complex interaction of many different factors. This seems to be the case with the brown lemmings of the Arctic tundra. If their young are born at the best time—at the start of the short summer—and if the summer is fine, large numbers of lemmings survive to spend the winter under the snow. Here, safe from predators, they go on breeding at a low level. If the next year is similarly beneficial, and if the winter weather is not too severe, huge numbers of brown lemmings are alive and

healthy by the end of the second winter. There may be a 200-fold increase on the original population; each litter, born after only three weeks' gestation, contains four to eight young, which can in turn begin to breed after only five weeks. At high density, though, the lemmings start to destroy much of the vegetation, exposing themselves to predators—foxes, weasels, owls, skuas, and hawks—which themselves increase in numbers because of the abundant lemming prey available. Lemming migrations may occur under these conditions, but these do not in themselves control the population. It appears to collapse of its own accord as hundreds of non-migratory lemmings die or otherwise fail to breed.

The reasons for the crash are not fully understood. Food shortages may be important, but the collapse can occur when reasonable amounts are still available. Disease becomes significant in the crowded conditions, but a major effect of the crowding seems to be hormonal disorders resulting from stress. These factors all seem to act together, along with migration and predation, to cut the population down to a point where only a handful of the rodents survive the third winter to start the cycle again. At the same time, the predator populations shrink, too, as their food supply diminishes.

These lemming cycles last an average of four years. There is probably a connection between this and the relative infrequency of good summer breeding conditions. In general, the Arctic tundras enjoy only two good summers—necessary for the population build-up of lemmings—in any four-year period.

Such wild population fluctuations, though common in extreme environments, rarely occur in tropical rain forests. Animal populations there tend to vary relatively little from year to year. It was once thought that this was due to the complexity of the tropical forest ecosystem—that the complexity brought many checks and balances into play on populations. Recent research suggests that this is not true in every case. The fact seems to be, quite simply, that the food supply for almost any species in a tropical forest remains so stable that there is no real chance for vast and sudden changes.

A case in point is that of the populations of small fruit- and insect-eating birds in the rain forests of Sarawak. Although the forest produces abundant food the birds get a limited share of it, because a large animal population exploits this food all the year round. We saw in Chapter 5 that each of the many insect species is present only in low numbers, and many have evolved camouflage mechanisms that make them hard to find. Moreover, there is no such phenomenon as the very great increase of available plant material that occurs in the spring in ecosystems such as the tundra and the deciduous forest nearer the poles. It is this increase that allows tundra animals, such as lemmings and migratory birds, to rear large numbers of young.

There are seasonal changes in the Sarawak forests, with a small increase in insect numbers during and after the monsoon rains, which bring a peak of tree-leafing. But the supply of fruit varies only sporadically. All the small Sarawak forest birds rely on insects for feeding their young and most of them therefore breed at this season of comparative insect abundance. But because the increase is not very great, the insects are hard to find, and the amount of predation on nests is high, the forest birds manage to rear very few young. The insectivorous birds give their young intensive care, and remain with them for up to six months after fledging until they are able to find their scarce and camouflaged food. There is not time then for producing second broods, as do many birds in the colder regions. On an average, each small bird pair in the Sarawak forests manages to rear only 0.34 offspring to juvenile age. These few juveniles then have difficulty staying in the area of forest where they hatched, for the adults engage in territorial behavior just after the young have been reared, thus excluding most of the young from the breeding population. They can readily gain a territory only if an adult dies, but the adults are long-lived and suffer little predation. Furthermore, the dry season immediately follows the period of territorial display. Insect food is now in shorter supply, with many different animals competing for it. The earlier territorial behavior of adult birds has apparently ensured that, in this lean period, individual birds are sufficiently widely spaced to survive on the limited resources. Fruit-eaters suffer less in the lean period, although fruits too are sometimes relatively less abundant at this time. So it is that bird populations retain a year-to-year stability.

It seems, then, that the stability of environmental conditions in tropical rain forests contributes to maintaining stable animal populations, because the animals are almost always

Right: the population of tropical-forest insect-eating birds, such as this tailor bird of Asia, remains stable from year to year because the environment, and consequently the birds' food supply, shows only moderate fluctuations.

Below: the brown locust of South Africa is usually kept under control by the arid climate. But when favorable conditions occur, a population can build up to a level where it can rapidly destroy much of its own food supply.

Right: the volcano of Surtsey burst above the waves off Iceland in 1963, providing virgin rock surfaces for the later gradual colonization by plants.

Above: lichens and mosses are among the first plants to colonize bare rock. After years of growth they build up a soil in which successively larger and larger plants may take root.

Right: climax fir forest, the relatively stable end-point of plant succession in parts of temperate North America. About 10,000 years ago, when the great northern ice sheet retreated, this ground was left as bare of vegetation and soil as the newly emerged Surtsey.

living close to the limit of their food resources. By contrast, the great fluctuations in food availability in less stable environments, such as the Arctic tundra, can contribute toward extreme changes in the extent of populations. The actual size at which these populations are stable or around which they fluctuate may result from complex interactions of food abundance, predation, and the stresses of crowding. Ultimately, however, animal populations must be limited by the resources that are available to them.

Plant populations in an ecosystem also change

through time, but they tend to do so more slowly than animal populations. Individuals of many species of land plants other than annuals have longer life-spans than animals and are better able to resist short-term environmental fluctuations. But fluctuations between day and night and from season to season are not the only type of time-induced changes; there are also long-term changes in every ecosystem. One generation of forest trees provides food and shelter for many generations of animals. Over long periods of time, however, as climate changes, as new plant species evolve and animal populations alter, the whole ecosystem is transformed.

In the shorter term, when physical disturbance of an ecosystem's vegetation occurs—through land movements, perhaps, or through floods, storms, or some sort of human activity—regeneration to the earlier state usually takes place once the disturbance is over. Where a tree falls in a tropical forest or where a patch of farmland in once-wooded countryside is abandoned, unscarred woodland is likely to reappear, even though it may take up to 200 years for large trees to mature. This regeneration of vegetation involves a succession of different plant communities, as we saw earlier in Chapter 4. In moist, temperate areas of Europe and North America, for example, fast-growing, widely dispersing annuals (many of them grasses) are usually the first to spring up on cleared ground if the soil is still present. They are succeeded by perennials, which steal a march on annuals in the spring of the second year because they have reserves of stored food. After a few years shrubs and small trees become established. Then come large trees—conifers perhaps, and then the big broad-leaved species.

Where soil is absent and bare rock exposed, colonization is a slow process. The main feature of plant succession on a bare surface is that the colonizing plants—the first of which are generally lichens—must interact with their environment to produce new conditions that other plants can then exploit. Until generations of small plants have gradually built up a soil on bare rock, no trees can grow. And so the primary succession on rock takes much longer than the secondary succession in a forest clearing.

Plant succession characteristically ends with what is known as a *climax* community or formation—a type of vegetation that will thereafter alter only slowly. Such a climax community is the deciduous forest dominated by oaks, beeches, or maples that grows up in many well-watered parts of Europe and North America if farmland is abandoned. Once established, this kind of community will remain almost unchanged for hundreds of years. The plants are in equilibrium with their surroundings, and the dominant types of plant life are not subject to competition from larger species. Climax communities usually have a greater biomass and complexity than any of the successional stages that precede them, but their productivity is often lower.

Ecologists once believed that in particular environmental conditions on land, succession starting on a bare surface such as sand or rock, or in a pond or lake, would always follow the same series of stages. In other words, it was thought that one set of plant species would initially modify the environment, that it would be replaced by another, until the climax community appropriate to that situation was established. This was just a hypothesis, for the changes involved can take so many decades that they are difficult to follow. And recent studies, relying on techniques such as the analysis of plant parts and pollen preserved in mud, have suggested that the hypothesis was mistaken. We now believe that plant communities do not follow inevitable successional courses, and that the exact nature of successional vegetation depends a great deal on chance: upon what seeds happen to reach the right spot at the right time, and what changes happen to occur in the physical environment.

Eventually, a climax community characteristic of particular environmental conditions does tend to take hold, but its ancestral communities are not always the same. Vegetation certainly has an effect on the physical environment; it must do so, because so many materials pass between the two. The vegetation of a mature tropical forest, for instance, stabilizes the soil it grows on by slowing down the rate at which erosion occurs and acting as a buffer against the extremes of weather. Vegetation existing at any one time must be at least a partial factor in determining the type of species that will succeed it. This is especially true if the existent vegetation helps to establish or change the soil. Nevertheless, it is the nature of the physical environment that will ultimately determine what *can* grow.

Plant succession is not merely a matter of vegetation interacting with its physical environment, for animals have a strong effect on what happens. The types and abundance of animals that live on and around the plants can greatly influence both their successional stages and the final climax—just as the changing vegetation naturally affects changes in an animal community. Succession is not the result of some special intrinsic property of plant communities; it is just one aspect of the changing of whole ecosystems in the course of time.

Over long periods of time the earth has undergone fundamental changes of climate; for example, there were tropical swamps in the now-temperate zones in the distant past, and there were ice ages in the north in the quite recent past. Such changes are caused by two rather distinct processes, both involving relationships between sun and earth. First, during millions of years the continents, drifting gradually across the planet's surface, have been moving into different climatic zones. Fossils of plants and animals in Antarctica, for instance, show that it has not always had a polar climate; it has apparently drifted to its present position from much warmer latitudes to the north. Not only have the continents moved, but the position of the poles and therefore of the equator, have also changed through time. In both these ways the alignment of land to the sun has been altered.

The second major way in which the climate of a land mass may change is through fluctuations in the activity of the sun itself. Because the sun's radiant energy "drives" the earth's weather, any increase or decrease in the amount of energy transmitted to the earth's surface is bound to alter the climate; it was probably such a decrease that brought on the ice ages. And there are also shifts in the pattern of air currents in the atmosphere, linked with the variations in the sun's activity. Some such shift is the probable cause of the present drought along the southern fringe of the Sahara Desert. At the present time the Sahara seems to be expanding slowly southward. This is no new phenomenon. As a result of long-term climatic changes, the vegetation of Africa has undergone several immense transformations in the past. Let us take a look at what happened, not forgetting, of course, that the experience of Africa is interesting not because it is unique, but because it has been shared—in different ways—by all the continents:

These rock paintings found in the western Sahara give evidence of much wetter conditions that prevailed only 5000 years ago, when the now-desolate area must have been rich enough in pasture to support herds of cattle.

The ice ages of the Pleistocene epoch (which includes the last 2 million years of earth history) were not climatic events restricted only to the north polar regions. Glaciers also advanced and retreated in southern lands, and the climates of equatorial regions were affected. There was probably a worldwide drop in temperature during the ice ages, with sea levels falling as water became locked up into ice. These events would naturally have affected the pattern of air movements. The result for Africa has been a series of very great variations in temperature, rainfall, and evaporation rate. These have had enormous ecological consequences for the continent.

Neolithic rock paintings and remains show that woodland clothed the Ahagga (or Hoggar) Mountains of the central Sahara around 6000 years ago, and that deer—today found only on the northern fringe of Africa—browsed in the Tibesti Mountains, even farther south. People with domestic animals apparently traveled the Sahara widely: fish-harpoons that have been found resting in the dry bed of the Wadi Azouak in the Air (or Azbine) Mountains of the southern Sahara testify to the wetter conditions that once prevailed. Until some 8500 years ago, Lake Chad (now around 8000 square miles at high water) was as large as the Caspian Sea (around 164,000 square miles).

The eastern Sahara may well have been much drier than the western part in Neolithic times, however. And much earlier the desert may have been even more extensive than it is today. In a now-fertile part of western Africa, about 300 miles south of the present limit of shifting Saharan sand dunes, a line of old dunes has been traced beneath the overlying cultivation. These old dunes probably date back more than 22,000 years. Such a southward advance by the desert would have pushed all the West African vegetation belts toward the sea, and fragmented the coastal tropical rain forest.

Climatic changes in Africa have not been restricted to the north. Geologists and paleontologists no longer accept an old theory that periods of glacial advance in the north (some of which occurred as recently as 10,000 years ago) coincided exactly with increased rainfall in tropical Africa; but lake and pollen deposits do show that there have been great fluctuations in rain-

fall and in the extent of forest vegetation in eastern Africa in the last 15,000 years. Examination of the soil beneath much of the dense and humid Congo Basin forest shows that the forest is far from primeval: the trees are rooted in Kalahari sand, which dates back only 50,000 to 75,000 years ago, when the ecosystem of the area must have been not forest but semi-desert.

Just as the glaciers advanced from the poles and then retreated during the Pleistocene epoch, so have those on Africa's high peaks. Glaciers on the nearly 17,000-feet-high Ruwenzoris (Ptolemy's "Mountains of the Moon") now reach no lower than 14,000 feet, but at times during the Pleistocene they were much more extensive, pushing moraines down to an altitude of only 65000 feet. Such cool conditions must have affected the vegetation below the glaciers as well. With the advance of the glaciers, lowland rain forest must surely have been replaced over large areas by montane vegetation.

Whatever the exact pattern of climatic change in Africa during the Pleistocene it is clear that vegetation changes have been immense. Few African ecosystems can have existed in stable climatic conditions for very long periods. There has been an ecosystem succession through time, and this must have had important consequences for animal distribution and the evolution of species. Populations of African lowland rainforest animals must at times have been restricted to small isolated refuges of moist, warm climate. Then, at other times, they have had a potential home stretching from the Atlantic to the Indian Ocean. Similar changes, as we have said, have occurred elsewhere in the world. Eurasia and North America, subjected to several great invasions and retreats by blankets of ice, have been particularly affected by Pleistocene events. In Southeast Asia, the rain forests must have been fragmented and rejoined several times, as sea levels wavered.

Thus, the Pleistocene seems to have been an epoch especially marked by change. Change is restricted to no time or place, though; plants and animals live in a world of ecosystems that has always been changing and that will continue to change in the future—from minute to minute, from day to day, from season to season, and from millennium to millennium.

The Lewis Glacier on Mount Kenya. During the last 2 million years, glaciers on tropical African mountains have advanced and retreated by thousands of feet, radically changing the distribution of plant and animal communities on the lowlands beneath.

The Distribution of Species

The fragmentation of a tropical rain-forest ecosystem through climatic change can readily bring about the origin of new species. When a rain forest is broken up in this way, its fragments become essentially islands in a sea of other vegetation. No two of these island refuges will be exactly the same. The soils and terrain beneath them will be different, as well as the weather above. They may not all start with exactly the same set of plants and animals, either. As each forest island is different, it puts different evolutionary pressures on the species living within its confines. These species become adapted to the particular conditions in their "islands"; thus, many of the species found in the original unfragmented rain forest will produce a whole set of new species—one to each forest island.

If, after a long period of time, the climate changes again, restoring the original conditions and allowing renewed rain forest to grow up between the fragments, the different species with a common ancestor may have become so distinct that their populations occupy slightly different ecological niches and can therefore live together when they meet. A degree of competition among them will lead to even further separation of their niches. This process seems to have been an important factor, for example, in producing the many species of monkey that inhabit the tropical forests of Africa.

Geographical isolation has probably been the major factor behind the origin of most species. But fragmentation of a population is not the only way in which geographical isolation can come about. It can happen because of the activities of species themselves. For instance, part of a population of plants or animals may invade an isolated area—an island in the ocean, a mountain top, or a cave—and having arrived there may become adapted to the peculiar conditions of the place, thus taking up different ways from those of the ancestral population. This is what happened in the Galápagos Islands where species have invaded from the South American mainland 600 miles to the east; once there, finding few of the competitors that were present on the mainland, they have become adapted to a whole range of new ecological niches.

The Galápagos finches are a famous example of this process—one of the examples, in fact, that

sparked Charles Darwin's imagination and contributed to his theory of evolution by natural selection. One invading finch from America has produced 13 new species in the islands, with beaks adapted to different diets. Members of one species even act like woodpeckers, probing crevices for insect larvae with cactus spines held in their beaks. Until the finches evolved to fill it, the woodpecker niche on the Galápagos Islands had been vacant.

A population cannot become a distinct species unless it is somehow isolated from its ancestors: otherwise, interbreeding will keep it genetically similar to them. But new species do arise for other reasons than ecosystem fragmentation or invasion of isolated areas. As time passes, an ecosystem changes even if it is not fragmented, and species living there must become adapted to the new conditions. For instance, when an area is invaded by a new species, those already in the system must become adapted to this fresh ecological situation: the new species inevitably has effects throughout the ecosystem web. Just as a single species may in time become several as a result of geographical isolation of populations, so a single species may evolve into another that is different because of changes in its surroundings. Both processes are essentially the same: isolated populations becoming adapted to new niches in changed ecosystems.

The myriads of different species that have been produced by spreading populations and changing ecosystems live within very different sorts of geographical ranges. The Devils Hole pupfish, for instance, is the most restricted of vertebrates, for it can be found in only one rock-girt spring in the American State of Nevada, close to Death Valley. The Texas blind salamander lives in only a handful of caves and wells on the Balcones Escarpment in Texas. And there are a great many other species that occur in only one lake or on one tiny mid-ocean island. By contrast, the moorhen (or common gallinule) is one of the most widespread vertebrates. It is found throughout

Red crabs, marine iguanas, and blue-footed boobies occupy distinct horizontal zones on the shore of the Galápagos island of Fernandina, illustrating both large-scale and local aspects of distribution. A limited number of species have adapted to specialized niches on this and other islands in the group.

most of Europe and much of southern Asia, in temperate and tropical America, and in Africa south of the Sahara. The wren of Europe occurs right across Asia and into North America, where it is known as the winter wren. Between these extremes are a great number of different ranging patterns: some species occur throughout a major ecosystem type in one continent, others in several ecosystems, and others only in one limited part of a major ecosystem. And then, too, there are the migrants, particularly birds, which have different geographical ranges at different times of the year.

The difference between the pupfish and the moorhen is that the former has adapted to an ecological niche with a limited geographical range, the latter to a niche with a very widespread range. In the Nevada spring, a niche for a small alga-browsing fish was exploited by the pupfish's ancestor. But the spring is an extremely limited habitat unconnected by surface water to any other aquatic environment. An animal specialized for life there is not adapted to life on the dry surface outside. It might do well in some similar freshwater habitat in other parts of America or the world; but it does not have the means of dispersal to get there.

The moorhen, on the other hand, is adapted to life in swamps and at the edges of any inland waters that have thick shore vegetation, in which it nests and on which it mostly feeds. It is equally at home swimming in water and walking on land, and it can fly too. Over the moorhen's wide geographical range, the type of water, climate, and species of plant in the environment vary greatly; but the differences in environ-

mental factors are less marked than those between spring water and dry land in Nevada.

The moorhen, too, is an adaptable bird, with very broad environmental tolerance. It feeds on a wide range of plant and animal material, and its populations can disperse quickly over wide areas. Thus, it is much more ecologically flexible than the pupfish or the blind salamander. Such specialist species are more vulnerable to changes in the ecosystem than more adaptable forms.

Because of their narrow requirements and limited distribution, there are more specialist species than adaptable species in the living web. The moorhen fills one niche over a wide area, but a number of different species of blind salamander are found in caves over the same area. For instance, the Texas blind salamander and the European olm are both amphibians with tails, but they belong to quite different families. Both, however, live underwater in underground caverns and show similar adaptations to this life. The skin of the adults lacks pigment, their eyes do not function, and they breathe with gills instead of lungs. This is an example of *convergent* evolution —two different species in different parts of the world evolving very similar appearances as a result of adapting to the same kind of ecological niche. Such evolution occurs where, when one potential niche is common but widely separated, a single species is unable to disperse to fill the niche in each place.

Examples of convergence are found in most groups of plants and animals. In the humid paramos, above the timberline of the high equatorial Andes, plants known as *Espeletia* (closely related to the sunflowers) have evolved into bizarre

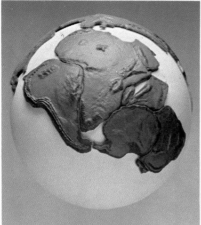

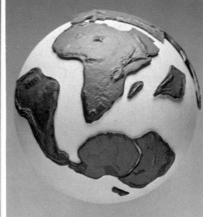

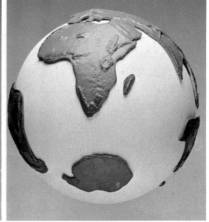

150 million years ago, the continents of the world formed a single landmass.

75 million years ago, Australia, South America, and Antarctica were joined.

50 million years ago, continents began moving to their present positions.

The three globes, above, illustrate three of the stages in the breakup of the old Southern Hemisphere landmass of Gondwanaland, which began about 135 million years ago.

gigantic forms like great cabbages perched on small tree trunks. This is the result of their adaptation to the year-round conditions of intense nighttime cold and intense daytime illumination, with a great deal of ultraviolet light. Far away in the African equatorial mountains, groundsels, quite different members of the same family, have converged to fill the same sort of niche. These groundsels (called *Dendrosenecio*) occur in giant treelike form in the high-altitude African heathlands.

In rocky areas at similar heights in these same African mountains live hyraxes—small, chubby animals (about the size of large rabbits) with short ears, a very short tail, and short legs. Hyraxes form a distinct order of mammals not closely related to any other, although possibly they had a distant ancestor in common with the elephants. The rock hyraxes live in colonies among tumbled rocks, feeding mostly by day on grasses and other vegetation and frequently sunbathing on rock surfaces. One of their typical communicating calls is a high-pitched whistling noise, which warns other hyraxes of approaching danger. They are widely distributed throughout the dry areas of Africa and western Asia and are found at altitudes of over 14,000 feet on mountains such as Mount Kenya. In the Ruwenzori Mountains, on the eastern edge of the Congo Basin, however, the high-altitude, rock-dwelling hyrax is a different species: the tree hyrax. The Ruwenzoris are girdled by moist forests, and the rock hyrax has been unable to pass through

these to reach the high heathlands. So the tree hyrax has filled the niche that the rock hyrax normally takes. Elsewhere, the tree hyrax lives in the rain forest, climbs trees, and is nocturnal instead of diurnal. But in the Ruwenzoris, it behaves precisely like the rock hyrax: instead of climbing trees, it lives among rocks at 12,000 to 14,000 feet; and instead of feeding only at night, it is active during the day. It even has a high-pitched whistle as an alarm call.

In the mountainous areas in other parts of the world, unrelated small mammals occupy a similar niche to that of the African rock hyrax. Marmots (also known as ground squirrels) in northern Eurasia and mountain viscachas (rodents related to the chinchilla) in the Andes are examples of just such rock-dwellers.

As we have seen, isolated populations of animals and plants that occupy similar niches in similar but isolated ecosystems may be totally unrelated species that have evolved a similar way of life (such as the hyrax and the marmot); or they may be members of the same family though different species (like the rock and the tree hyrax). In the dry grasslands of South America, Africa, and Australia (another widespread but highly fragmented type of ecosystem) live a number of large, long-legged flightless birds—the South American rhea, the African ostrich, and the Australian emu. Although the appearance, habits, and diets of these birds are all very much alike, many zoologists used to think that they were not closely related, but had

A flock of ostriches in South West Africa.

Male South American rhea with chicks.

Male Australian emu with chicks.

The present distribution of ostriches, rheas, and emus may have resulted from continental drift. The ancestor of these birds probably inhabited a single landmass that broke up millions of years ago into Africa, Antarctica, Australia, India, and South America.

converged in isolation to occupy similar niches. Modern research, however, indicates that probably they, together with cassowaries, kiwis, tinamous, and the extinct moas and elephant birds, are related. They are collectively known as ratites. The ancestral ratite apparently lived in the great southern continent of Gondwanaland, which began to break up about 135 million years ago. The emus, rheas, and ostriches probably had a South American ancestor. And we now believe that some of its descendants spread into Australia by way of Antarctica before the three continents drifted apart about 50 million years ago; and others were isolated in Africa when it was divided from South America 90 million years ago.

The drifting of continents can clearly have considerable influence on the distribution of species, not only by carrying land into different climatic zones but by physically separating or bringing together populations. Other earth movements, such as mountain-building (which may also be caused by the movements of continents), can also split up populations. These movements, combined with climatic and other changes, and with the varying specialization and powers of dispersal of animals and plants, have led to the present pattern of species distribution.

Adaptable Man

One species has spread into more ecosystems than any other. It has named itself *Homo sapiens*—"wise man"—and the name describes the nature of man's tremendous adaptability. Physical structure is the major feature of most species' adaptation to their particular ecological niches. The most widespread species are in general the most highly adaptable *physically*. But man, the most widespread of all animals, has been successful not so much because of his anatomy—though that has been important—but because of his intelligence.

Zoologists do not regard man as especially distinctive physically. He is classified as one of 150 living species of primates. The primates are a group of mammals (hairy, backboned animals that maintain a constant warm body temperature and feed their young on milk).

According to recent studies, man probably diverged from the apes about 15 million years ago, in the Miocene epoch. A manlike animal then lived in the forests of Africa and India and a few fossils of this creature still survive; he has been given the name of *Ramapithecus*. He probably looked and lived rather like a modern chimpanzee, but he was smaller, with small canine teeth and shorter muzzle. It is possible that *Ramapithecus* made tools and, like chimpanzees, included some meat in his diet. Like most living monkeys and apes, he almost certainly lived in social groups. Gradually, more manlike primates evolved from this man-ape.

In the succeeding Pliocene epoch (which lasted from about 7 to 2 million years ago), the climate of Africa apparently became drier. As during the dry phases of the later Pleistocene, rain forests probably gave way to savanna grassland over wide areas. The ancestors of chimpanzees and gorillas stayed in the fragmented forest, but man's ancestors became adapted to existence in the grasslands. As they evolved truly bipedal

Men inhabited the wooded savannas of Africa 4 million years ago. The tool-using and hunting species Australopithecus africanus *probably lived in cooperative groups of about one dozen. In this artist's reconstruction, a group of these early people are pictured going about their daily tasks near the rock overhang under which they have passed the night.*

The Bushmen of the Kalahari Desert, like Australopithecus africanus *of 4 million years ago, hunt wild animals, and gather plants instead of practicing agriculture. Their survival depends on an intimate knowledge of both the living and the non-living environment.*

locomotion, walking and running upright on their hind legs, their hands were freed for other functions; and hunting probably became a much more important part of their lives.

After *Ramapithecus*, the evolutionary story remains vague until about 5 million years ago. Fossils of a true early man of the African savannas from this later period have survived: *Australopithecus*, the "southern ape." There were at least two species of *Australopithecus* in the savannas. One, more gracefully built than the other, was perhaps a direct ancestor of modern man; we call him *Australopithecus africanus*, and he was a chimpanzee-sized bipedal animal, probably living in cooperative bands on the savanna, sheltering at night in cave mouths, existing on an omnivorous diet (of which about 20 per cent was meat), and using simple stone tools. Bipedalism and the use of weapons made him a wide-ranging and potent predator, although anatomically he was no match for the big carnivorous mammals, and ecologically he was only one consumer competing with many others in one small corner of the biosphere.

Very recent discoveries in the fossilized shoreside deposits around Lake Rudolf in northern Kenya suggest that before the beginning of the Pleistocene epoch 2 million years ago, man in Africa had evolved even further toward the modern human condition than had *Australopithecus africanus*. By then there existed a species of manlike animal that, although small in size, was so nearly like us that he should probably be classed as *Homo*.

By one million years ago, men whose size and shape were much like ours were in existence. They spread out from Africa into Asia; their remains have been found in Java and near Peking. These men—given the name of *Homo erectus*—were larger than any of their predecessors, and with bigger brains (though not quite as big as modern man's), shorter faces, and heavy brow ridges. *Homo erectus* used fire to release stored solar energy from plant tissue, and he manufactured large, fairly efficient tools. Since the days of *Australopithecus* he must also have developed language well beyond the limited range of sounds that apes combine with gestures in order to communicate with their fellows.

The large, intelligent, articulate, tool-making and fire-using men of the middle Pleistocene began to have an enormous impact on their environment. They differed greatly from other omnivores, such as bears, in that they did not rely on their anatomical structure and on stereotyped behavior patterns to get food and defend themselves from enemies. Man was beginning to mold his environment to suit himself, instead of being subject to all its vagaries. By hunting cooperatively and intelligently, and by using both fire and weapons as an aid to frighten, trap, and kill game, he was putting enormous new pressures on prey animal populations; thus he probably contributed to the extinction of some large mammals. But the crash of a prey population did not lead to the collapse of human populations too, for adaptable man was capable of making a switch to other foods.

Males and females in the social group of middle-Pleistocene man were more nearly alike in size and structure than in most other primate species, and the young remained dependent on their parents for many years. During this long period of dependency, they learned the language of the group and, through the language, the complex techniques of tool-making, hunting, and shelter-building, and the rules of the society.

By 400,000 years ago, man's physical structure had evolved to a point where it was very close to that of modern man's. *Homo sapiens* had arrived. In the last half-million years, man's interactions with his environment and his social life have undergone immense changes, but without any great accompanying anatomical modifications. Learning has been of primary importance. Faced with new situations, man has been able to adapt to them by means of his intelligence.

Early *Homo sapiens* spread to all land areas of the world except the frozen wastes of Antarctica. But his diet was still little changed. As hunter and gatherer, he learned to eat particular items according to the ecosystems in which he found himself. In some areas, especially in cold climates, he was forced mainly to hunt; but in general, small items of vegetable origin formed the bulk of human food.

To the corners of the earth, then, man took with him his knowledge of fire, and his techniques for making stone tools, hunting weapons, and shelters. He also carried along the arts of painting and music, rituals governing the life of the community, and myths about its surroundings. All these attributes are cultural, not genetic. The use of fire and the manufacture of clothes and shelters, coupled with his dietary adaptability, allowed early man to survive in ecosystems where no other primate could live. Meanwhile, boat-building permitted him to

An aerial view of New York City from above the Pan Am building, Park Avenue. Man has evolved culturally to a stage where he creates his own ecosystems from which most of the ecological processes of the natural world have been erased.

disperse to areas inaccessible to other primates.

This life of the hunter-gatherer, with a small band of people (composed of a few related families) living in a large territory into which other groups may not trespass, still persists in such remote places as the deserts of Australia and southern Africa; but it is a disappearing way of life. Man's very efficiency as a hunter may have depleted many natural food sources; 10,000 years ago, he began to grow plants for food—that is, to manage their ecology, rather than relying on the usual processes of the ecosystem. Clearing the ground of natural vegetation, he planted the

Left: clearing and tilling the ground using hand tools, often after having first removed the forest, is still the main agricultural technique over much of Africa. Such a way of life, based on a small village, differs little from that of Neolithic man.

Below: goats and sheep at a watering place in Jordan. These two closely related hoofed mammals were probably among the first animals domesticated by man for their milk, meat, hides, and wool. Thousands of years of sheep and goat grazing have helped destroy the original vegetation around the Mediterranean.

tubers or seeds of food plants, especially grasses, and he harvested the swollen tubers or ripe fruits at the end of the growing season. This was the birth of agriculture, in what is known as the Neolithic (or New Stone) Age.

Agricultural practices depend for their success on one of the features of plant succession: that productivity is highest in the early successional stages. Annual and perennial herbs have a higher rate of production than do the larger plants—often trees—that form climax communities. They are also easier to harvest, and they produce concentrated packages of food in very short spaces of time. But to keep agriculture going the succession must be constantly pushed back by "weeding," plowing, and resowing, or by some other form of clearance to an early point. After a time, if such activities are continued, frequent harvesting of crops starves the ecosystem of its essential nutrient minerals, because these minerals have been taken up from the soil by the plants and then completely lost when the plants are harvested. When this happens, the agricultural community has three options: it can move on to a new site—as with the "slash and burn" method of cultivating rain forests—or it can allow fields to lie fallow in rotation, or it can constantly refertilize the soil. Many small Neolithic communities shifting from one place to another were not enormously different from hunting-gathering groups. Indeed some degree of hunting was probably still an important aspect of life among most Neolithic men. In some cases, however, hunting probably took on as much ritual as nutritional significance, because agriculture became more and more important. But more settled Neolithic communities, which were learning how to use the soil without entirely impoverishing it,

These terraced rice fields on the Indonesian island of Bali are typical of the intensive agriculture possible by using a basically Neolithic adaptation to life in the river valleys in warm climates. The first civilizations, where large numbers of people lived a close-knit existence, were able to develop once such irrigation techniques had been learned.

The ruins of the Inca city Machu Picchu, surrounded by its terraced cultivation plots high in the Peruvian Andes. The Inca civilization had reached a Bronze Age state of development when the Spanish Conquistadors arrived in the early 1500s.

tended to be much larger and to develop greater division of labor than the hunting band. The agricultural village became the customary social unit; and this has persisted up to the present day in many parts of the world.

In addition to cultivating wild plants and selectively breeding them in order to increase their yield and ease of harvesting, Neolithic man domesticated animals. These, too, he bred selectively, to bring out such desirable features as docility and cooperativeness in hunting companions, pack animals, and providers of food and clothing, as well as to obtain increased yields of meat, milk, and wool. The animals that man domesticated most successfully were those that lived naturally in social groups and had a built-in tendency to cooperate and to depend on a leader animal. The dog may have been domesticated from wolves even before the Neolithic Age.

It was followed by sheep, goats, cattle, pigs, and donkeys, and later by horses, camels, and llamas.

The most successful attempts at agriculture occurred in river valleys in warm climates. The rivers, which carried minerals down from their drainage areas, not only provided essential water but also constantly refertilized the soil. The warmth gave long, favorable growing seasons. When well-managed, the river valleys allowed very dense human populations to exist in one place over long periods of time. Under such conditions, diet became dominated by the cereal and other plant crops that sprouted from the rich soil. A new social order appeared, often with a strongly authoritarian, centralized organization, which could effectively manage a river valley with a complicated irrigation system. The chief of a band or settlement was superseded by a king and his officers. The city-based state appeared, with the city controlling a large area of surrounding territory, whose resources provided for the city's needs.

By now only one part of the population was needed to grow the crops and to care for the domesticated animals. And therefore many people were released for other, more specialized tasks. They began to smelt for the production of metals—soft bronze at first—from which they were able to make more sophisticated tools and weapons. Soon, too, they started to erect permanent buildings, to write, and to use money. In the valleys of the Nile, Tigris-Euphrates, Indus, and Yellow rivers, our civilization was born.

Natural ecosystems were greatly affected, of course, even by early agriculture and the domestication of animals. Vegetation around the village was cleared to make way for crops, and domestic animals destroyed or modified the vegetation wherever they grazed and browsed. Many trees were cut down for use in building, or as fuel to give warmth, to cook food, and to make pottery. Such Neolithic activities may have combined with the drying climate to turn the western Sahara into an infertile desert. Still, the effects of the Neolithic man's way of life were restricted

A modern industrial society may still have close links with its Neolithic past. Here a tanker discharges its oil cargo to a refinery in a rural corner of Britain, watched by sheep on a farm that has much in common with farms of thousands of years ago.

mostly to the immediate vicinity of each human settlement. Later civilization, however, had truly profound effects on the whole biosphere. As civilizations developed with growing momentum, they demanded resources from well outside nearby cultivated areas. Metal ores were mined far away; forests were felled to provide fuel for smelting and other energy-demanding processes; rocks were quarried for city buildings; special materials were gathered for making paper, ink, furniture, clothing, decorations, and weapons. Especially in royal households, the demand grew for exotic foods and strange things both animate and inanimate. Men began to travel far and wide in search of items that would give pleasure and entertainment. Some citizens were able to spend their time studying the cycles of sun and moon, the movements of stars, and the changing patterns of climate, and they constructed models to explain the nature of the universe. Mathematics and science had begun.

Civilization, which relied on travel, on the collection of materials from far and wide, and on trade with other cities and communities, inevitably spread to places outside its river-valley cradles. Then, as different civilizations vied for land and resources, wars flared up and disease occasionally decimated dense city populations. At times, war and disease acted as brakes on the development of civilizations, but these civilizations continued to grow and spread their influence across the world. East and west came into contact, and explorers and traders began to make regular long sea voyages, disturbing the

Below: flensing a fin whale (removing its blubber) at a whaling station. Whaling, like fishing, is one method of hunting that still has economic significance in the industrialized countries of the world, but whale meat is little used today as human food, and most of it goes for pet food. Modern hunting techniques have caused huge drops in several whale species.

Neolithic cultures and early civilizations of the Americas and Africa. Steel and gunpowder, printing, new crops, new trading patterns—innovations of every kind multiplied. Eventually, there came an enormous boost to growth and change when 18th-century England began to use fossil fuels to provide energy for manufacturing.

For many thousands of years charcoal (which is basically just charred wood) had been the primary fuel of craftsmen, but wood was growing scarce as forests close to civilizations were fast disappearing. Fossil fuels are very concentrated energy sources. Coal, especially after conversion to coke, allowed much larger-scale manufacturing processes to take place. Men built factories near the coal mines, and towns grew up near the factories. The Industrial Revolution was now under way. Since the 18th century, gas, oil, hydroelectric power, and nuclear energy have been added as industrial fuels to support industrial processes that have led to new technologies that demand even greater resources of energy.

Industrialization has led to radical changes in human ecology. In heavily industrialized countries, more men live in cities and work in industry (as well as in the commerce that industry supports) than live in agricultural areas. They are living not only on the daily local supply of energy and materials but also on accumulations from past eras, and the production of other lands. Our use of large amounts of energy, and of the machines and techniques invented by our societies, has made it easier for us to destroy natural ecosystems in order to provide more food and living space. Advanced technologies have made the killing of animals for food and sport easy and efficient. The rate of change has been accelerated by the improvement of long-distance communications and the spread of education. We have combed almost all the land surface of the earth and much of the sea in a vast hunt for resources to supply the ever-growing industries and their energy-consuming products. Every ecosystem has been exposed to the effects of industrial civilization.

Many ecosystems that have not been directly

Right: applying weedkiller to intensively cultivated land in the Colorado River Valley, southern California. Chemicals from the residues of some weedkillers can build up in the soil, and, when concentrated in large amounts in agricultural produce, may cause damage to developing embryos in animals and man.

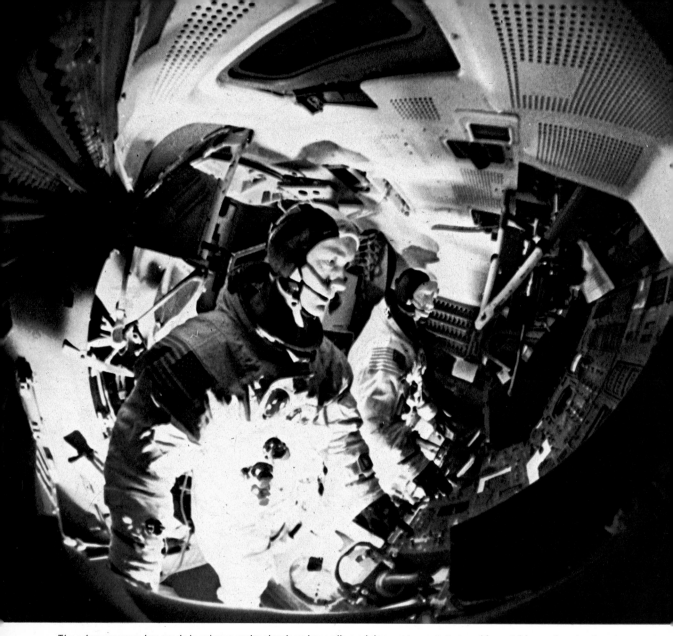

The advances man has made in science and technology have allowed these astronauts to travel far outside man's natural environment, but they have also put grave pressures on the natural systems that still support all life on earth.

"industrialized" have been grossly modified by industrialization elsewhere. For example, we have felled tropical rain forests, have planted large areas of the tropics with commercial crops that will not grow in temperate lands, have ranched southern grasslands to provide meat for the industrial north, and have raided the polar seas for seals and whales. Industrial man has left nearly untouched only the most inhospitable and unproductive polar ice caps, deserts, and mountain peaks; but even in these areas he has set up communities supported by energy, agricultural produce, and manufactured goods from elsewhere. The nature of human society itself has

changed, and there is a tendency for community life to be replaced by "nuclear families."

In the long-industrialized countries, few pieces of land have remained unaffected by man. We have covered huge areas with an ecosystem extremely different from that originally there, the artificial urban ecosystem of city, town, suburb, and industrial complex—an ecosystem entirely dominated by one species of consumer, and with the normal cycles of nutrients and flow of energy completely obliterated. We farm other areas intensively, in a style derived from Neolithic agriculture but highly modified by mechanization and industrial and scientific techniques. We often strip the natural vegetation from the remaining portions of less productive land by felling trees and opening it up to grazing. In some places, we now produce timber by planting whole forests of fast-growing, often exotic, trees. In others, we alter river courses, flood valleys, dig canals, and strew railways and roads across the land. Once-dominant plant and animal species have been pushed into small pockets where the remains of original ecosystems survive; elsewhere, man and his few cultivated and domesticated species have replaced them. Only a few particularly highly adaptable species have survived in large numbers in the human systems—species that we usually regard as pests.

Man-made productive ecosystems—the "countryside"—are much less stable than natural ecosystems. They must be constantly refertilized as nutrient minerals are removed along with the plant and animal crops. In the narrowed-down simplicity of their ecology, they are subject to the sort of "plagues" of herbivores and animal parasites typical of other relatively simple systems, such as the tundra. Such artificial ecosystems usually allow a higher rate of soil erosion than would the original vegetation.

To control crop pests and parasites, man has poured poisons into his artificial ecosystems, and these, persistent through many years and concentrated through food chains into top carnivores, have had powerful effects far removed in space and time from their point of application. DDT, for instance, has been found in the Antarctic snow! For centuries, industry and its related machines have emptied their wastes into the environment, poisoning the air, land, and rivers and affecting far-off ecosystems. Fertilizers draining from the land have upset the balance of river ecosystems. Much of the detritus of industrial civilization is carried out to sea. The sea is also threatened by one of the prime fuels of our present civilization, oil, which often escapes during transportation or extraction, posing a serious threat to marine life.

Meanwhile, man himself thrives in a way that creates additional problems. Mechanized farming and the influence of science on agriculture and medicine have increased the productivity of man's artificial ecosystems, and allowed many more people to survive to reproductive age and beyond. In Neolithic society, and to some extent in nonindustrial civilizations, starvation (brought about by crop failure or pestilence), natural disasters, disease, warfare, and social customs all helped to limit the size of human populations. Bands of hunters and gatherers must have been even more susceptible to environmental factors affecting their survival and food supply. Our modern world—free of most killing infectious diseases, producing food by intensive land management, cushioned against natural disasters, and released from many social taboos—has undergone a population explosion. Man has become one worldwide interbreeding population, which is pressed dangerously close to the ceiling of its resources.

We now realize that the fossil fuels that have supported the growth of industrialization will not last forever—and, indeed are likely to be exhausted quite soon if present growth continues. Despite our increasing ability to improve crop yields, the production of plants cannot possibly keep pace with growing populations.

The ecosystem that modern man operates in is the whole biosphere. He is a major consumer at most points of the food chain in many natural and artificial ecosystems within the biosphere, and he occupies a vast array of niches. He can exploit energy sources besides the sun and other living things in order to maintain himself in inhospitable environments and to circulate materials through ecosystems, but he is ultimately dependent on these systems for his continued existence. We all still rely for our food on green plants, which use sun energy for their production; and this production and water remain our most basic requirements. Unless we take account of the principles on which the biosphere functions, and unless we keep our population and consumption in balance with our resources, we cannot survive. Unless we protect and conserve large areas of natural ecosystem, the earth will lose a wealth of intriguing species that cannot be replaced; and this loss may have serious repercussions throughout the web of which we and they are a part. If we replace too much of earth's natural vegetation with very simplified agricultural ecosystems, the dynamic balance of the biosphere may be totally upset.

Man has been the first species to produce his own self-contained ecosystems so successfully that he can travel out of the biosphere. In the future he may even be able to travel outside the solar system. But he will not do so if he fails to succeed in managing his own population and its balance with the resources of Spaceship Earth.

Index

Page numbers in *italics* refer to illustrations or captions to illustrations.

leka keppe, 60, 62
lemmings, 112, 113–4
lesser flamingos, 32, *32*
Lewis Glacier, Mount Kenya, *121*
l'Hoest's monkey, 88, *90–1*, 92
lichen, 75, *116*, 117
life: history of, *12–13*; on earth, origins, 10
light: calorie value, 49; daily, effect on ecosystems, 104
limestone, 47; upwelling of, 49
lion, *36*, *40–1*; territorial behavior, 62
locusts, browsing behavior, *23*
luring devices of plants, 28

Macaques, grooming, *66*
machines, development of, 14
Machu Picchu ruins, Peru, *134*
mammals: first, 35; high-vegetation-eating, 25; migratory, 112
man: adaptable, 128–39; as competitor, 62; beginnings of civilized, 134; dependence on other organisms, 10; future survival, 139; in food chains, 41; interference in competition of species, 60; origins, 128, *128*
manatees, 32
marine iguanas, *122*
marmots, 126
marsupials, competition from other mammals, in Australia, 62
mating, 14, 66
merganser, *76–7*
mesquite tree, root system, 80
migrants and migration, 110–2, 124; effect on local ecosystems, 111
mineral salts, 47; in aquatic ecosystems, 96; in leaching, 47–9; oceanic, 16; upwelling of, 49
Miocene era, 128
mites, soil, 45
molecules, original, 10, *10*, 13–14
mollusks, 38
monitor lizard, *82*
monkey: "language" of, 73; of African rain forest, 88, 90–2; social groups, 71–2, *71*
moorhen, ecological flexibility, 122, 124–5
moose, *54*
mosses, colonization of bare rock, *116*, 117
mountain-building, 127
mushrooms, 127
musk oxen, *54*

Natural selection, 64–5
nectar and nectar-feeders, *26*, *27*, 28

Neolithic man, 133–5, *133*
nerve cells, *15*
New York City, *131*
nitrates, *46*, 47
nitrogen, 21, 47; cycle, *46*, 47
nocturnal animal life, 105, 106–8
nucleic acids, 14

Oak tree, energy flow, *52–3*
oceans: classification by environmental divisions, 92–3; stability, 95
offspring, numbers of, and parental cooperation, 68, 71
oil, 47; pollution of the sea, 139
olm, European, 125
organic matter: produced by deserts, 80; yearly production by photosynthesis, 23
oryx antelopes, *82*
osmosis, 95
ostriches, *127*
otter, *76–7*
over-grazing of Mediterranean lands, *132*
oxpecker birds, 75
oxygen, of water, 47

Padas River, Borneo, *84*
parasites, *39*, 41
parental care of young, 67; in early man, 131; in insectivorous birds, 114; in terns, *67*
peat, 47
Penicillium notatum, 60
permafrost of the Tundra, 111
photosynthesis, 16, 18, 21, 47; effect on day/night alternation, 104; energy trapped by, 49; in marine plants, 96; in water, 21
phytoplankton, 16, *16–17*, 21, 30, 96
piapiac birds, *73*
pike, *100*
plankton-eaters, marine, 36–7
plants: aquatic, 32; evolution of, *12–13*; first, 33; food of man, 132–5; of desert, 80, 82; of running water, *94*, 95–6; of Southeast Asian forests, 84; parts of, and uses, 25; population, changes in ecosystem, 117–21; protective devices, 30; successive communities, in regeneration, 117–8; *see also* phytoplankton
Pleistocene epoch, 87, 121, 130–1
Pliocene epoch, 128
pollenization, *26*, *27*, 28, 66
pond as ecosystem, *76–7*
population: control, in animals, 112; explosion, of man, 139; stability of animal, 114, 117; stability of plant, 117–21

porpoises, 37
Potomogeton, *94*
predators: and prey, 32–3, *40–1*: deep-sea fish, 99; detection methods, 35; early man, 130–1; in population control, 112–3; insect, of the desert, 82; inshore, 38; size factor, 38; vertebrate, 35
primary consumers, 23, 30, 36
primary producers, 23, *23*, 56, 58
primates, number of living species, 128
production by tropical rain forest, 84
protective devices of plants, 30
protozoans, 45
pupfish of Devils Hole, 122, 124

Radula, 27
rainfall: causing leaching, 47–8; determining desert, 80; in history of Africa, 121; of tropical-rain-forest belt, 83–4
Ramapithecus, 128
"raw materials" of living things, 47
red colobus monkey, 88, *90–1*
red crabs, *122*
Red Sea, salt content, 95
redtail monkey, 88, *90–1*
regeneration of vegetation, 117–8
reindeer migration, 112
remote control mating, 65, 66
reproduction, 14; effect of climate, 67
reptiles: body temperature, *54*; desert, 82; first, 33
respiration, 47; in heat production, 50–1
rhea, *127*
rhinoceros carrying oxpeckers, 75
rice fields of Bali, terraced, *133*
rivers: ecosystem, 93; primary productivity, 96, 99; salt content, 95; valleys, in early agriculture, 134–5
rock: formation from shells, 47; paintings in western Sahara, *118*, *121*; weathering of, 48
rodents, desert, 82
root-eaters, 28, 30, *30*
roots, 21; inhibiting secretions, 58; of cacti, 80
Ruwenzori Mountains, 121, *125*, 126

Saguaro cactus, 80, *105*
Sahara Desert, *81*: evidence of wetter origins, *118*, 121; shift of, 118
salamander, blind, of Texas, 122, 125
salt: content of oceans, 95; dif-

Picture Credits

Key to position of picture on page: (B) bottom, (C) center, (L) left, (R) right, (T) top; hence (BR) bottom right, (CL) center left etc.

Cover: H. Eisenbeiss
Frank W. Lane
2 Popperfoto
5 *Life* © Time Inc. 1974, J. Dominis, 1967
9 *Life* © Time Inc. 1974, (1969)
10(T) Josef Muench
10(B) Icelandic Photo © Mats Wibe Lund Jr.
11(R) Josef Muench
14(TL) University College, London/Photo Michael Freeman © Aldus Books
15(TR,B) © Miss Gene Cox, Micro Colour Ltd.
19 John Oates
20(L) L. Lee Rue/Bruce Coleman Ltd.
20(TR) John Markham/Bruce Coleman Ltd.
20(BR) © Douglas P. Wilson
22, 23(L) Simon Trevor/Bruce Coleman Ltd.
23(C) Jane Burton/Bruce Coleman Ltd.
23(R) John Oates
24(T) David Hughes/Bruce Coleman Ltd.
24(BL) Spectrum Colour Library
24(BR) S. C. Bisserot/Bruce Coleman Ltd.
26 Cyril Laubscher/Bruce Coleman Ltd.
27(L) Heather Angel
27(R) Bellieud/Pitch
28 Jane Burton/Bruce Coleman Ltd.
29(T) John Markham/Bruce Coleman Ltd.
29(B) Graham Pizzey/N.H.P.A.
30 Ron Boardman
31 Norman Myers/Bruce Coleman Ltd.
32 Heather Angel
33 L. R. Dawson/Bruce Coleman Ltd.
34 Zig Leszczynski/Animals Animals, New York, © 1974
35(T) Cyril Laubscher/Bruce Coleman Ltd.
35(B) Popperfoto
36(T) *Life* © Time Inc. 1974, G. Silk, 1961
37 Rabrt/Jacana
38 G. J. H. Moon/Frank W. Lane
39(T) Heather Angel
39(BL) Peter Hill, A.R.P.S.
39(CB) Heather Angel
39(BR) © Miss Gene Cox, Micro Colour Ltd.
43 *Life* © Time Inc. 1974, J. Dominis, 1967
44(TL) Jane Burton/Bruce Coleman Ltd.
44(BL) Edward S. Ross
45 U.S. Navy photo
51 G. R. Roberts, Nelson, New Zealand
54(T) Ardea
54(BL) Photo Michael Freeman © Aldus Books
55 L. Lee Rue/Bruce Coleman Ltd.
57 Douglas Fisher
58–9 Heather Angel
61 Peter Scoones/Seaphot
62 L. Lee Rue/Bruce Coleman Ltd.
63(T) © Douglas P. Wilson
64 Francisco Erize/Bruce Coleman Ltd.
65(R) W. L. N. Tickell
66 G. J. H. Moon/Frank W. Lane
67 *Life* © Time Inc. 1974, A. Eisenstaedt, 1965
68(L) Sester/Pitch
69 Emil Schulthess
70 *Life* © Time Inc. 1974, A. Eisenstaedt, 1965
71 Treat Davidson/Frank W. Lane
73 Francisco Erize/Bruce Coleman Ltd.
74 Jeff Meyer/Animals Animals, New York, © 1974
79 NASA photo
81 Emil Schulthess
82 Eliott/Jacana
83(R) H. Beste/Ardea
85 J. Alex Langley/Aspect
86 Photo Michael Freeman © Aldus Books
87 J. L. Mason/Ardea
89(TL) Jane Burton/Bruce Coleman Ltd.
89(TR) Des and Jen Bartlett/Bruce Coleman Ltd.
89(BL) Bruce Coleman Ltd.
89(BR) S. C. Bisserot/Bruce Coleman Ltd.
92 Prevost/Jacana
93(R) ZEFA/G. Koehler
94 Heather Angel
95 © Douglas P. Wilson
97 Popperfoto
98(T) Jane Burton/Bruce Coleman Ltd.
98(B) © Douglas P. Wilson
100(T) Jane Burton/Bruce Coleman Ltd.
100(B) Des and Jen Bartlett/Bruce Coleman Ltd.
101 Anthony and Elizabeth Bomford
103 Trans-Antarctic Expedition Committee
104 Frank W. Lane
105 Lewis Wayne Walker
106–7 Russ Kinne/Bruce Coleman Ltd.
108(B) Hans Reinhard/Bruce Coleman Ltd.
109 S. C. Bisserot/Bruce Coleman Ltd.
110(L) Joseph Van Wormer/Bruce Coleman Ltd.
111 *Life* © Time Inc. 1974, D. Kessel
113 C. J. Orr/Bruce Coleman Ltd.
115(T) Cyril Laubscher/Bruce Coleman Ltd.
115(B) Mike Holmes/Animals Animals, New York, © 1974
116(T) Solarfilma, Reykjavik
116(BL) Dennis Brokaw
117 Dennis Brokaw
119 Mme Irène Lhote
121 Gerald Cubitt
123 Heather Angel
124 J. L. Mason/Ardea
125(TR) John Oates
125(BR) G. Manson-Bahr
126 Photos Michael Freeman © Aldus Books
127(T) C. Haagner/Bruce Coleman Inc.
127(BL) Jane Burton/Bruce Coleman Ltd.
127(BR) V. Serventy/Bruce Coleman Inc.
130(L) Simon Trevor/Bruce Coleman Ltd.
130(R) Photo Laurence K. Marshall
131 Howard Sochurek/The John Hillelson Agency
132(T) Larry Burrows/Aspect
132(B) Photo Eric Hosking
133 J. Alex Langley/Aspect
134 Victor Englebert/Susan Griggs Agency
135 The Studio Jon Ltd.
136 R. W. Vaughan/Bruce Coleman Ltd.
137 Georg Gerster/The John Hillelson Agency
138 Photri/Aspect

Artist Credits

© Aldus Books: Gino D'Achille 128–9; David Nockels 12–3, 40–1, 76–7, 90–1; Peter Bakelin Associates 46, 48, 52–3; Peter Warner 16–7

THE WEB OF LIFE

Part 2
Invisible World

by Derek Toomer and
Alan Cane

Series Coordinator	Geoffrey Rogers
Art Director	Frank Fry
Design Consultant	Guenther Radtke
Editorial Consultant	Donald Berwick
Series Consultant	Malcolm Ross-Macdonald
Editor	Allyson Rodway
Copy Editors	Maureen Cartwright
	Damian Grint
Research	Enid Moore
	Peggy Jones
Art Assistants	Vivienne Field
	Michael Turner

Contents: Part 2

Editorial Advisers

DAVID ATTENBOROUGH. Naturalist and Broadcaster.

MICHAEL BOORER, B.SC. Author, Lecturer, and Broadcaster.

MATTHEW BRENNAN, ED.D. Director, Brentree Environmental Center, Professor of Conservation Education, Pennsylvania State University.

PHYLLIS BUSCH, ED.D. Author, Science Teacher, and Consultant in Environmental Education.

MICHAEL HASSELL, B.A., M.A. (OXON), D.PHIL. Lecturer in Ecology, Imperial College, London.

STUART MCNEILL, B.SC., PH.D. Lecturer in Ecology, Imperial College, London.

JAMES OLIVER, PH.D. Director of the New York Aquarium, former Director of the American Museum of Natural History, former Director of the New York Zoological Park, formerly Professor of Zoology, University of Florida.

Introduction

In our urbanized society, many people give hardly a thought to the earth's wildlife or, in fact, to such domesticated organisms as cattle and corn. Our plant and animal foods come prepacked, and we can easily forget the life behind the wrappings. So it is not surprising that we are even less conscious of the existence of a world of living creatures too small to be seen (except perhaps when they erupt as fungus or gather together in vast colonies). If we think of microscopic life forms at all, we tend to characterize them generally as "germs," for among them are some disease-bearing types of bacterium and virus, as well as fungi and algae.

Yet life on this planet would not be possible without the microorganisms. They are essential to the breaking down of dead organic matter so that it can return to the air, sea, and soil; they play a vital role in providing nutrition for plants and in helping herbivorous animals to digest their food; without them we could have neither bread nor cheese; and many of them keep our bodies healthy in strange and wondrous ways even though others do, indeed, cause illness.

From humanity's lofty vantage point, these minute organisms appear to be the lowest forms of life. But we cannot deny that in an evolutionary context they are the most successful of living creatures. They have exploited every possible habitat, including some that would seem highly improbable. Their very simplicity makes them extraordinarily adaptable, able to meet the challenge of changing environments. And so they survive and multiply while more sophisticated life forms stumble toward extinction. It is probable, in fact, that the despised microbes will inherit the earth when all the rest of us have gone. Their invisible world, then, is worth close study, and this volume may well induce some readers to delve more deeply into it.

The Unseen Universe

Over the course of millions of years, the living organisms best adapted to the conditions in which they lived have survived and produced descendants; those less well adapted have perished. This inexorable process of natural selection has given rise to the distinctive shapes, forms, and ways of life of all the living things, plant and animal, that we see around us—and to an entire universe of living creatures that are much too tiny to be seen by our unaided eye. Indeed, until powerful magnifying glasses and microscopes were developed in the 17th century, the existence of these strange microorganisms (or microbes, for short) was unknown, and natural clues to their existence—disease, decay,

Familiar sights of land, sea, and air—trees, seaweed, sunset clouds—belong to the world that we know. Over, under, in, and on them live countless tiny creatures whose importance far surpasses their size. See pages 10–11 for a first glimpse of some of the "invisible" inhabitants of these three scenes.

and alcoholic fermentation—were attributed to magic or divine intervention.

It is with these remarkable microorganisms that this book deals. Despite their size, they are among the most important and interesting of living creatures, with an influence on the earth and its inhabitants that is far out of proportion to their dimensions. Among them are invisible creatures that have drastically changed the course of human history. For example, the bubonic plague, once known as the Black Death, which eliminated a quarter of the population of Europe during the Middle Ages, is caused by a microorganism, a bacterium that we call *Pasteurella pestis*. Other microbes produce a variety of poisonous substances: tetanus, for example, is caused by a powerful poison produced by a bacterium called *Clostridium tetani,* and another bacterium, *Clostridium botulinum,* causes a terrible kind of food poisoning called botulism, which may result in paralysis and death.

But although these examples suggest that microorganisms are chiefly harmful, this is far from the truth. Most of them are beneficial to other living creatures. There is, for instance, a whole range of microorganisms whose activities replenish and enrich the life environment, and on which all living creatures depend ultimately for their survival.

We shall be looking at a few organisms that

we already know well. Mushrooms and toadstools, for example, are the visible clues to the existence of certain kinds of fungus, the other parts of which are invisible to the naked eye. For the most part, however, our tour through the world of microorganisms will be a new and fascinating experience. That is why the names we must use for identifying and describing them are strange. They cannot have common names like "mouse" or "buttercup," because they can be seen only with the aid of a microscope, and so there has never been any need for describing them in familiar terms—apart, that is, from talking vaguely about "germs." The names that identify them are names that make sense primarily to scientists, because scientists are the people who do see and deal with them.

Scientists of all nations use the same scientific

Fertile land has a vast microbial population; in one ounce of good earth there are millions of amoebas like the one below (left). The ocean's plant life appears to be composed mainly of kelp and other large seaweeds—unless, under a microscope, we view the myriad

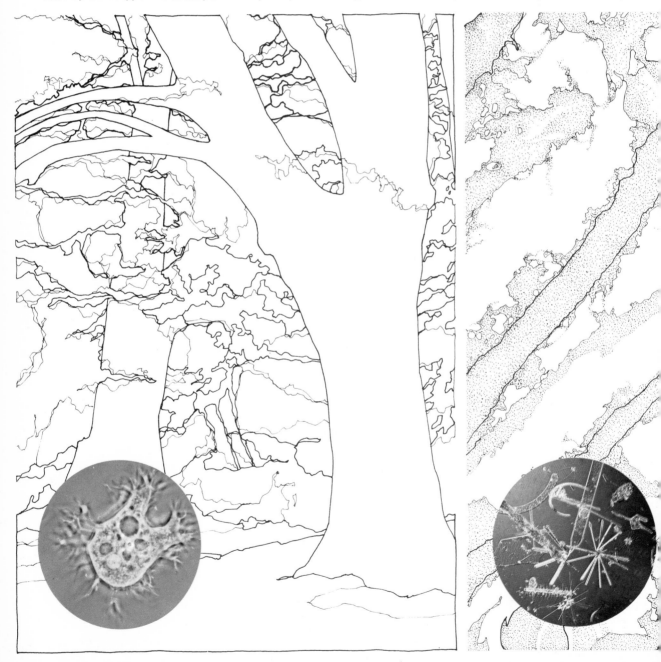

names for all plants and animals. This is for two simple and practical reasons. First, they want to be sure they are talking about the same organisms, and secondly, they choose names that indicate relationships between the many different sorts of creature.

The names of microorganisms are really less intimidating than they look, and it is perfectly possible to appreciate the complexity and excitement of the invisible world without knowing the precise names of all its inhabitants. But it is interesting to know how scientists refer to living organisms—especially because we shall occasionally be using a technical term or two in these pages. Each name consists of two parts. For example, in *Clostridium botulinum*, the food-spoilage organism, *Clostridium* is the generic name: it is used for a number of organisms

algae that make up the phytoplankton (center). Liberated spores of the fungus Puccinia graminis, *seen (right) germinating on a barberry leaf, can be blown hundreds of miles. "Clean" air is often alive with various microorganisms.*

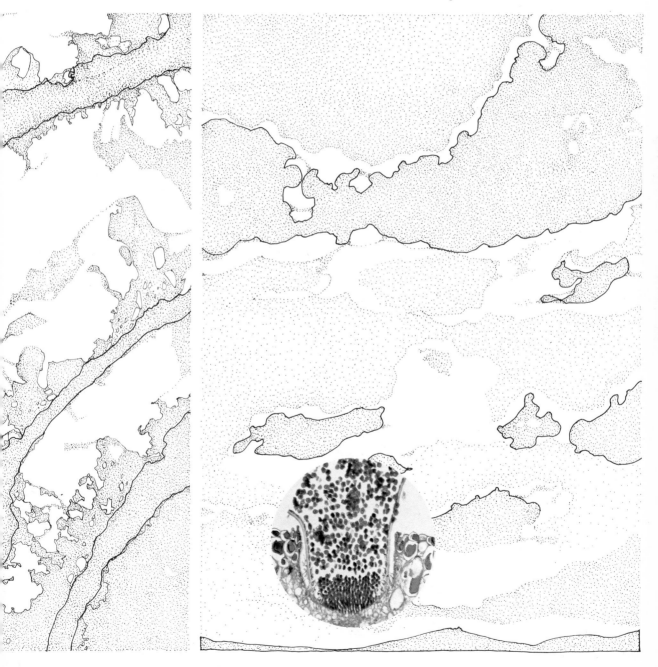

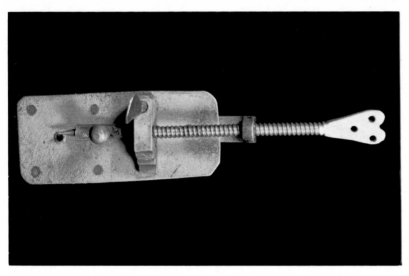

The above ancestor of the electron microscope below has a lens embedded in a metal base; by turning the screw, a viewer could shift an object impaled on its point into viewing position. Primitive? Perhaps. Yet it produced 17th-century drawings such as these at right of two molds, one growing on leather (top), the other on a rose leaf.

all very much alike but with small distinguishable differences. The generic name is often abbreviated to its capital letter alone. The second name, *botulinum,* refers to a particular kind of *Clostridium* organism. Others are *C. tetani, C. perfringens,* and so on. The same pattern holds for the other creatures of the invisible world (as well as for all visible organisms), with the single exception of the viruses. These—the tiniest and strangest of living things—are often identified simply by code numbers such as PLT22 or T4.

Some scientists believe that microorganisms are the most successful of all living things. It may seem obvious that man, with his ability to find technological answers to most of the problems posed by the climate and geography of the world, is the outstanding success story in the history of this planet. Microorganisms as a group, however, are immensely successful because of their adaptability and variety. The history of antibiotics illustrates how superbly microorganisms adapt in order to overcome adverse changes in their environment. Antibiotics such as penicillin are chemical substances that are extremely effective at killing microorganisms; yet within a few years of the discovery of penicillin and its subsequent widespread use by physicians, there appeared new, penicillin-resistant microbes that caused the same diseases. No matter what antibiotic or combination of antibiotics doctors use, microbes resistant to that antibiotic appear after a time. Even more remarkable is the fact that some microorganisms that acquire resistance to an antibiotic are able to transfer this resistance to other species. That is one reason why the search for new and more effective antibiotics is a chief and extremely important preoccupation of the *pharmaceutical,* or drug-producing, industry.

Man has been able to colonize inhospitable areas of the earth chiefly because of technical and social skill. And like other mammals he can maintain his own body temperature in a wide range of conditions, and the various solutions such as blood, lymph, and so on, that bathe the organs inside his body are kept at just the right consistency. In extreme conditions he constructs defenses to protect himself. Microbes have no sophisticated technical skills, nor can they maintain a constant internal "environment." Yet there is no place on earth that man has colonized where they do not also live. And they inhabit places where neither man nor any other animal or plant could survive.

Because microbes as a whole are so various, they do not form a tight, easily defined group of creatures like, say, the mammals or the fishes, with features in common that distinguish them from other groups. All that the microorganisms have in common is their comparatively simple structure and small size. Whereas higher plants and animals are constructed out of many thousands of cells all working together, a microorganism is only a single cell, or at most a number of single cells joined together in a colony.

The cell is the fundamental unit of which all living organisms are made; and although cells come in many shapes and sizes, every cell is constructed on roughly the same pattern. It is basically a minute blob of jellylike material separated from the outside world by an enveloping membrane, and containing a control center, or *nucleus.* The nucleus not only controls the cell's activities, but also contains the hereditary material that determines what sort of cell it is and what it does.

Many of the cells in higher plants and animals are specialized for particular functions—producing digestive juices, for example, or carrying nerve impulses. Microbes differ in that all their activities are carried out within a single cell or colony of similarly functioning cells. So each microbial cell is, in a sense, equivalent to the entire complicated assemblage of cells that make up the bodies of higher organisms.

In our summary of the invisible world, we shall be looking at five main kinds of microorganism: the protozoans, algae, fungi, bacteria, and viruses. Most members of these groups fit our criteria for inclusion in the invisible world—simplicity of structure and small size—but there are exceptions. We shall not, for example, discuss the larger algae (the seaweeds), fascinating though they are, for they are hardly *micro* organisms; some, indeed, can reach a length of over 100 feet. So we shall concern ourselves only with smaller kinds of algae.

Actually, the word "small" covers a multitude of sizes, for although no microorganism can be seen clearly with the unaided eye, there is a great difference in size between the smallest microorganism and the largest. Because they are all so tiny it is difficult to compare them with anything familiar, and conventional measuring techniques are meaningless. So scientists

Amoeba (Protozoan)

Pennate diatom
(Alga)

Chlorella (Alga)

Paramecium (Protozoan)

A Comparison of Sizes

(If the amoeba were as big as
a house, the other microorganisms
would be approximately the size
of the objects shown.)

Life-Forms of the Invisible World

Penicillium notatum (Fungus)

Yeast (Fungus)

Salmonella typhi (Bacterium)

Staphylococcus aureus
(Bacterium)

T4 Bacteriophage (Virus)

These drawings are based on photomicrographs of 2 specimens of each of 5 major groups of microbe: protozoan, alga, fungus, bacterium, virus. As shown in the comparative-size sketches of familiar objects, size differences among microbes are great, ranging from the amoeba's "gigantic" 500 μm (500 millionths of a meter) down to the "tiny" polio virus at 0.03μm.

Poliomyelitis virus

use a special unit for measuring such minute creatures; it is called the micrometer and is written thus: μm. One μm is a millionth part of one meter.

The smallest of all living organisms are the viruses, which may be only 0.03 μm in diameter. Other microorganisms are much larger. Bacteria are usually about one μm in diameter, and the fine threads of fungi may measure as much as five μm across. The giants of the microbial world are the protozoans, some of which, at 500 μm, are just visible as fine specks to the naked eye. Algae come in many different sizes, but the single-celled forms such as *Chlorella* are about 10 μm in diameter.

Few of the microorganisms that we shall be meeting in this book can be classed as either plants or animals, but the algae are clearly true plants, whereas the protozoans are for the most part clearly animals. Plant cells differ from animal cells in that plant cells are surrounded by a rigid wall made of a chemical called *cellulose*. The chief distinction, however, between plants and animals is the fact that green plants make their own food, using the energy of the sun to convert water and carbon dioxide gas into sugar. This process, called *photosynthesis,* is brought about by the pigment chlorophyll, which is found in all green plants. Algae make their food in this way. Protozoans are animal-like in that they must hunt for food. A watery environment is essential to the survival of both.

Protozoans are a very diverse group of organisms. They may live in or on other living creatures, or they may be free-living—that is, they may fend for themselves in nature. There are more than 30,000 known species of these single-celled creatures, most of which have some form of locomotion to help them find food. The different ways in which they move through water serve to distinguish different kinds of protozoan. One group, for example, which

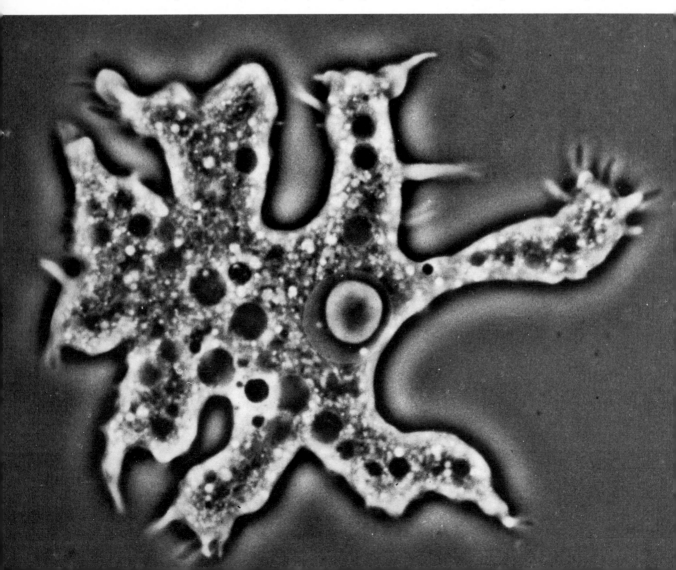

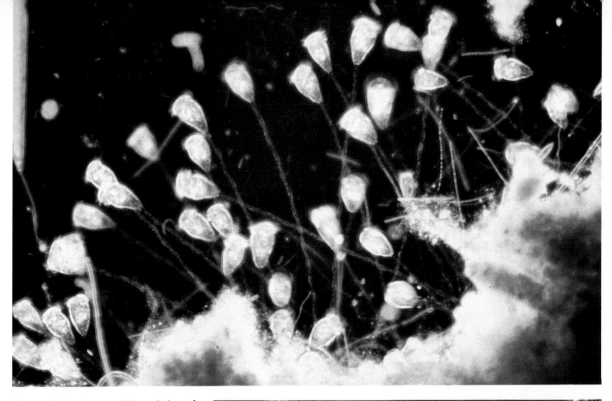

Three of the innumerable varieties of water-dwelling protozoan, which are animal-like in that they move about and hunt for food. An amoeba (left) sends out projections—pseudopodia—that change its shape and keep it flowing along. Vorticellas (above), with their bell-shaped bodies on stems, may resemble plants, but the head can leave its stalk and swim freely, assisted by vibrating cilia. And the long whiplike flagellum protruding from the body of each Euglena (right) helps to propel the organism through the water.

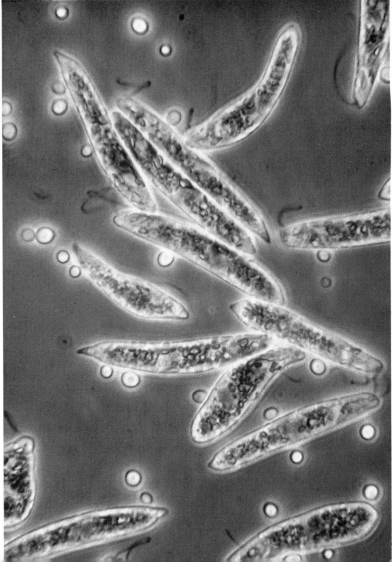

includes the protozoan that causes the disease amoebic dysentery, moves in a manner known as amoeboid movement. The best-known member of this group, *Amoeba proteus,* looks like an irregularly shaped mass of jelly, and moves by sending out flowing projections called *pseudopodia* that change the shape and position of the cell. If the pseudopodia form mainly in one direction the whole amoeba flows that way.

Other protozoans have more sophisticated ways of moving around. For instance, there is a whole group that use long whiplike beaters, known as *flagella,* to propel themselves through the water. Some members of this group—such as the protozoan that causes sleeping sickness in man—have a flagellum that runs the length of their elongated bodies and is attached to an undulating membrane that assists movement. Tiny beaters called *cilia* assist the movement of some protozoans. The cell surface of the slipper-shaped *Paramecium* is covered with cilia, which are much shorter than flagella. Paramecia are unusual in that they have a definite "mouth" region at one side of the cell, and the cilia around the mouth are especially modified to create a vortex in the water to pull food particles into the mouth as the paramecium cell moves along.

But not all protozoans have a means of propulsion. For example, for the major part of its life cycle the protozoan that causes malaria merely floats in the bloodstream, absorbing food through its cell wall. Some protozoans even build houses around their naked cells. The amoebalike creatures called *foraminiferans* secrete shells of chalk or silica, often of great beauty, around themselves. *Foraminifera* means "holebearers,' a name that refers to the numerous holes in their shells through which they send out long jellylike strands of protoplasm to capture food particles from around them.

Like the protozoans, algae need a watery habitat. They differ in that they are green plants and can create their food photosynthetically. The simplest of the many kinds of algae are free-living, single-celled organisms that move by thrashing two whiplike flagella. In this they resemble the free-living, single-celled protozoans, and, indeed, at this level of simplicity only the presence of chlorophyll serves to separate the plants from the animals. Why should algae have any need of movement, if they do not have to hunt for food? The answer is that they must "hunt" for sunlight.

In more complex algae, many flagella-equipped cells may be associated in a spherical colony. The flagella from *each* cell are directed to the outside of the sphere, and the colony rolls through the water as a result of their coordinated beating. Each cell in such a colony is, of course, microscopic in size; the colony itself is only the size of a pinhead. Many algae have no means of movement, especially those that consist of long filaments of cells joined end to end. *Spirogyra*—a colony of algae readily recognized because the chlorophyll is contained in a distinctive spiral-shaped structure inside each cell, like a green spring—is of this kind. One end of its filament is generally tethered to a rock or to another plant in quiet fresh water, and the rest of it floats.

The mainly free-floating algae known as *diatoms* are found worldwide—in seas, in fresh water, and in moist soil. The characteristic feature of a diatom is its hard siliceous cell wall, which surrounds the alga like a box. These hard walls are finely sculptured into remarkably beautiful patterns, but they are so small that it takes a microscope to reveal their full splendor.

Often considered with the algae, but not strictly related to them, are the blue-green algae. (The name comes from the combination of blue and green pigments in their cells.) These microorganisms, which include single-celled and filamentous forms, occur in a wide range of habitats, including extreme environmental situations such as saline lakes and desert soils. They are not classified with true algae because their cells are relatively simple in structure; they lack distinct nuclei, for example. They therefore form a separate group whose exact position in the evolutionary development of other microorganisms is still being investigated by microbiologists.

So far we have taken a preliminary look at organisms—the algae and the protozoans—that are at least recognizably akin to familiar plants and animals. Now let us meet some groups of microbes that are quite unlike anything we know in the visible world. The first such group comprises the invisible parts of the fungi. Fungi are rather like green plants in that their cells are

Though single-celled, the slipper-shaped Paramecium *has a very complex structure, the outer surface of which is covered with tiny beaters called cilia. These paramecia have been highly stained to show the nucleus (blue) and numerous cavities, or* vacuoles, *in which food particles (red) are enclosed for digestion.*

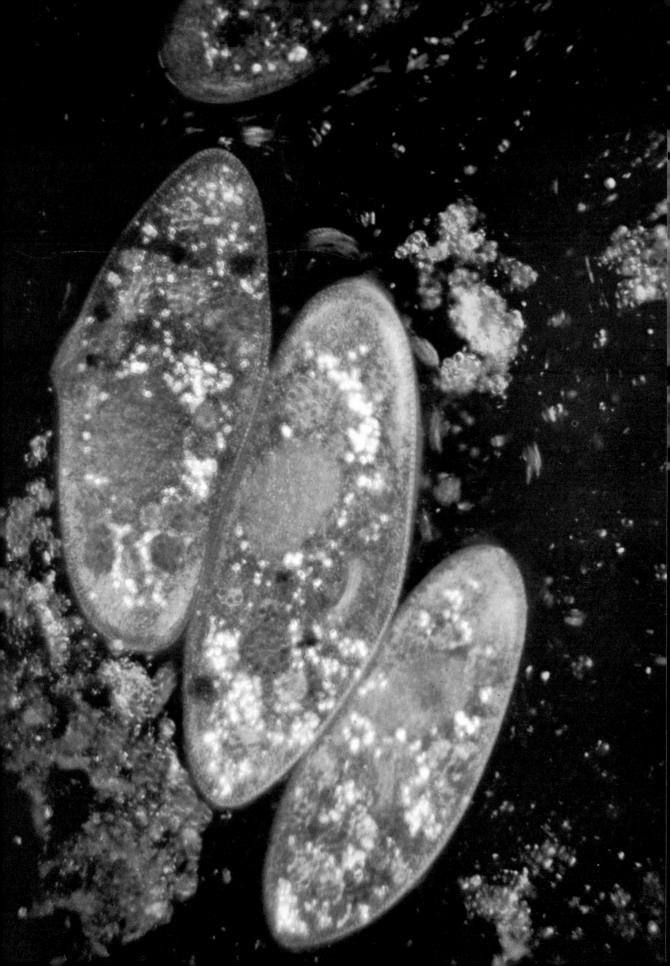

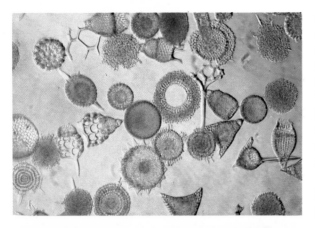

surrounded by a rigid cell wall; but whereas plant cell walls are stiffened with cellulose, fungal cell walls are stiffened with another chemical, chitin, which is similar to the hard substance that covers and protects the bodies of insects. The fungi are unlike familiar plants in that they have no chlorophyll and are thus unable to build up their own food substances from simple chemicals. Like animals, they must depend on ready-made food, whether dead or alive.

Thus the fungi are either *parasites* (that is, living in or on other living organisms) or *saprophytes* (living on dead or decaying organic

Right: the White Cliffs of Dover in England are composed of masses upon masses of microscopic shells—the incredibly varied and often lovely "houses" that the foraminiferan protozoans build around themselves. These shells sank to the bottom of the sea when the foraminifers died, and formed a thick layer; later changes in sea level exposed this to the air. As seen in these specimens (pink-stained in the top photograph, green-stained in the bottom one), the shells of the foraminifers, which are mostly marine creatures, are perforated with minute holes, through which protrude strands of protoplasm for capturing food.

matter). Many of the saprophytes, and some parasites, produce mushrooms and toadstools—the fruiting bodies that the fungus sends up above the ground to disperse its *spores* (reproductive cells that will give rise to new individuals). The main body of the fungus consists of a fine network of branching colorless threads, or *hyphae,* that form a system called the *mycelium.* The mycelium, which grows on the substance from which the fungus draws its nourishment, is often invisible to the naked eye.

Parasitic fungi also produce mycelia, and these can eventually kill the host. One fungal parasite, for example, lives on the common housefly; its spores adhere to the insect's body, and on germination they sprout fine tubes that pierce the skin and bud off roundish cells inside its body, each of which develops into a mycelium. In time the fly dies, choked with fungal threads.

The housefly parasite belongs to the smallest and most primitive class of fungi. Among the 1000 or so species in this class are the water molds, which cause the fringes of colorless, threadlike hyphae seen on dead fish or other once-living material floating in lakes and canals. Also in the group are the parasitic fungi that are

responsible for the disease of downy mildew in many plants and the saprophytic fungi that make damp bread or fruit moldy.

The largest class of fungi, with almost 30,000 species, is distinguished by a special kind of spore-producing unit, the *ascus,* in which the spores develop in a long file before being liberated into the air. This group includes the fungi that produce the brilliant orange structures found on rotting wood or on the forest floor. It also includes the below-ground species *Tuber,* which produce the edible truffles highly prized by gourmets. And another member of the group, of immense importance to nearly everyone in the world, is yeast—or, rather, the various yeasts that are essential for brewing and baking.

The second largest fungal category includes about 13,000 species, among them all those saprophytes that send up fruiting bodies: the toadstools, bracket fungi, puffballs, and stinkhorns. Also in this class are many tiny fungi, including the dreaded food-crop parasites the rusts and smuts. All these fungi are identified collectively by the production of a characteristically shaped spore-producing structure, called the *basidium.*

The last group of fungi is actually composed of several kinds, which, because they seem to belong to no precise category, are lumped together under the catchall name of Fungi Imperfecti. Among these "imperfect" microorganisms are the fungi that cause such skin diseases in man as ringworm and athlete's foot. And there are also some microorganisms that, although not true fungi, are frequently included in the group: the slime molds. At certain stages a slime mold resembles nothing so much as a smear of white or colored jelly. Often to be found on fallen tree trunks and branches, slime molds feed on decaying plant matter such as rotting wood, leaf litter, and even old bracket fungi. It is rather like a mass of amoeboid creatures combined into one slimy sheet of jelly without any barrier membranes between the individual cells. The characteristic that slime molds have in common with the fungi is that they are able to produce complex and often colorful fruiting bodies to disperse spores.

We now turn to what is probably the most significant group of microorganisms: the bacteria. These are extremely simple, single-celled organisms, which may be spherical or oval, rod-shaped, filamentous (threadlike), spiral, or vibrioid (shaped like a comma). They may be found in chains, clusters, or pairs of cells.

The structure of the bacteria is simpler than that of any other group of organisms except the viruses, which are not even true cells. A bacterial cell, although it *is* a cell, is unlike, say, a protozoan or an alga in that it does not have a true nucleus. Each cell is surrounded by a relatively thick and rigid cell wall, which may have a coating of gelatinous material, the capsule, with a probably protective function. Many species have one or more flagella by which they move, and these may be concentrated at one end of the cell or arranged in a variety of ways.

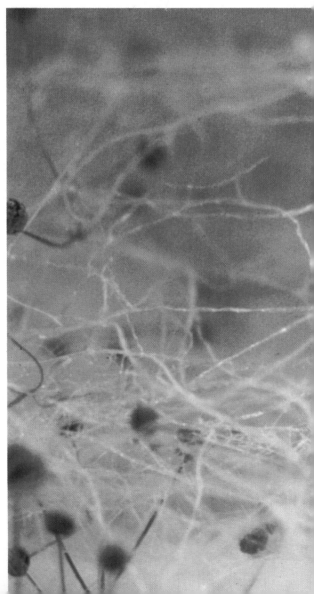

Flourishing on moldy bread is the cobwebby mycelium of the fungus Rhizopus *(common bread mold), with its spherical spore sacs rising into the air. Released spores of this fungus are everywhere, ready to germinate on damp organic material.*

Their structure is a poor aid to classification, and so to naming, although oval-shaped bacteria are appropriately called *cocci,* from the Greek word for "seeds," and rod-shaped bacteria are also appropriately termed *bacilli,* from the Latin for "little staffs." For the most part, however, characteristics other than shape and structure are used as guides to describing and naming the various species. Thus there are lactic-acid bacteria, which can ferment milk-sugar, and iron bacteria, which derive their energy by a chemical reaction involving iron; there are enteric bacteria, which live in the intestines of man and other animals; there are also gliding bacteria, photosynthetic bacteria, and so on. They can also be divided into two groups, Gram-negative and Gram-positive, depending on whether they can be stained with a dye developed by the Danish bacteriologist Christian Gram.

This rather condensed recital of the physical characteristics of bacteria gives no hint of the diverse nature of their activities or of their importance. The most fundamental and far-reaching characteristic of bacteria is their ability to survive and multiply.

No other microorganisms are as prolific as the bacteria, which multiply extremely rapidly by dividing. Some species reproduce themselves so quickly that a single bacterium can give rise to over a million offspring in a few hours. There is also a process of sexual reproduction; the "male" and "female" bacteria come together (the sexes are physically identical, but they play different roles in mating), and part of the hereditary material of the male passes to the female through a bridge of living material that develops between them. The two partners then separate, and both begin to divide. But only the cells derived from

the "female" bacterium show combinations of the characteristics inherited from both parents.

Bacteria do not produce spores for reproduction. But they sometimes produce endospores, which are spherical or oval structures inside the bacterial cell. Brought forth only in unfavorable conditions, these are the most resistant living things known. They can survive drying out, disinfectants, and extremely high temperatures.

It is appropriate here to look at the variety of reproduction methods in the other groups of microorganisms. With the exception of the viruses, all of them follow much the same pattern as the bacteria: they multiply by division.

Although many protozoans simply divide, there is a primitive kind of sexual reproduction in most species, which transfers hereditary material from one protozoan to another. In *Paramecium*,

for example, there are two nuclei, one large, one small, to each cell. The larger nucleus, whose job is chiefly to control the coordination of the free-living cell, disintegrates when one paramecium cell comes into close contact with another. The remaining nucleus in each of the paramecia divides into four, three of which degenerate, leaving only one. This divides into two, and one of these is exchanged between the partners. The two pieces of nucleus in each

Because they lack a supply of chlorophyll, fungi cannot manufacture their own food, but are forced to depend on external substances, whether dead or alive. Some, such as those below, are saprophytes: *that is, they eat decaying plant or animal matter—for example, the bark of a dead tree (left) or sheep's dung (right). Other fungi are parasitic; as their fungal threads branch out, they progressively infect the host, as in mildewed fruit (right), and perhaps even eventually kill it, as in the mycelium-choked housefly (far right).*

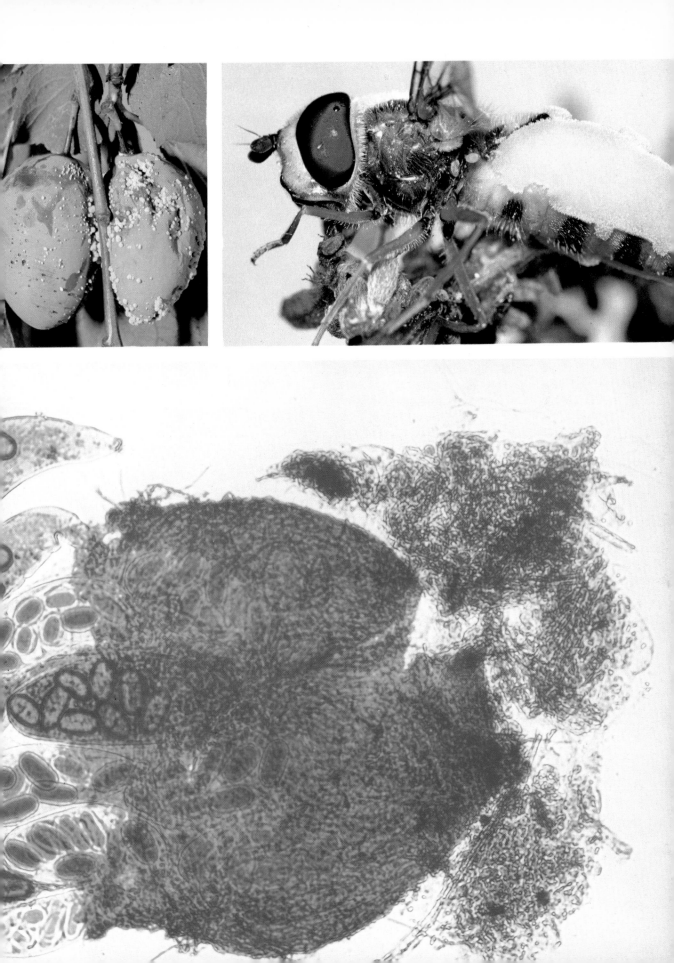

partner now fuse, and the paramecia separate. The new composite nucleus in each of them divides into four, and the paramecium itself divides in half, with two nuclei to each new cell. Thus four paramecia have been formed from the original two, with a mutual exchange of hereditary material. Some protozoans also form spores—roughly spherical structures with hard, resistant walls—which can remain alive during bad conditions and give rise to new protozoans when conditions improve.

In the algae and fungi, multiplication by division of a single cell is only one of a number of ways of reproduction. Division can also occur through the cutting off of a new part of the filament in a filamentous alga or fungus. In most of these microorganisms, though, some kind of sexual reproduction takes place. The spores produced in the asci and basidia of the fungi that possess such structures are the result of sexual reproduction.

With the viruses, we come to the smallest and some of the simplest of all living things, and also to the strangest type of microbial reproduction. Carl Linnaeus, the 18th-century Swedish biologist who devised the system of scientific names that we now use to classify living things, described three natural kingdoms, animal, plant, and mineral. If he had known about viruses, he would have been hard-pressed to decide which of his three kingdoms they should be assigned to. They live only in other living organisms—in plants and animals, and some, the so-called *bacteriophages*, in bacteria. "Live" is perhaps a rather questionable word, because they exhibit none of the usual properties that we think of as denoting life.

The whole existence of the viruses is devoted to the production of more viruses. To do this, they invade living cells and ultimately destroy them. Once inside a cell, they take control of its nucleus and divert it from its normal function to the task of manufacturing and assembling new viruses. They cause a range of diseases of varying degrees of severity; in human beings, for instance, viral infection brings on such complaints as the common cold and smallpox.

Viruses have been described as "living chemicals," and they are often, in fact, barely distinguishable from mere chemicals. Some of the smaller ones—for example, the one that causes mosaic disease of the leaf in tobacco plants—can be crystallized from solution in the same way as common salt crystals. But the virus loses little of its potency through such drastic treatment. If dissolved again and injected into a healthy tobacco plant, it will again cause the same mosaic disease.

Despite their extreme simplicity, these mysterious organisms have a distinct structure, which the superior power of the electron microscope reveals. Some, for example, are geometrical structures called *icosahedrons*—that is, they have 20 triangular faces and 12 corners. Bacteriophages, the viruses that live on bacteria,

The slime molds, though not actually fungi, resemble the fungi in their ability to produce fruiting bodies for the dispersal of their spores. To the naked eye, the masses of single-celled creatures that combine into visible slime molds look like smears of white or colored jelly as they flow along decaying plant matter such as damp logs or leaf litter. Under the microscope, however, we get a clear view of their spore cases (technically called sporangia), as in these colorful photographs of green, yellow, and brown varieties of slime mold.

27

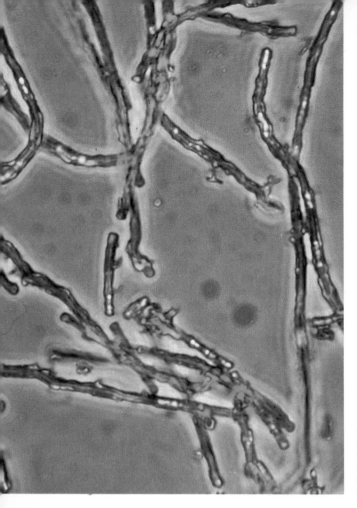

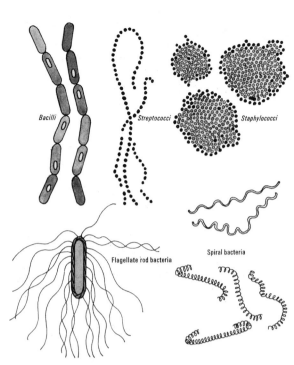

Above: five of the many types of bacterium, which are often named according to their shapes: the bacilli, for example, are rod-shaped (Latin bacillum *means "little staff"), and the cocci are spherical (from the Greek for "seed"). Left: when these anthrax bacilli attack cattle or sheep, the infection is usually fatal, and it can even be transmitted to man.*

have a head, a tail, and an irregular array of tail fibers. Some viruses, such as the tobacco mosaic virus, appear as long rods with a coiled spring-like center. Whatever their shape, however, all of them are little more than bundles of hereditary material surrounded by a coat of protein.

Now that we have seen something of the kinds of creatures that inhabit the invisible world, we shall now discover something of their activities. Only in comparatively recent times have microscopes been powerful enough to show us these tiny creatures in action; but some of the shrewder early biologists suspected that something more than supernatural powers was involved in phenomena such as the souring of milk and the onset of disease. In the 1600s, lens-making techniques were sufficiently advanced to allow Anton van Leeuwenhoek, a Dutchman, to make the first investigation of microbial life. It was a profound experience for him. "What if one should tell people in future that there are more animals living in the scum of the teeth in a man's mouth than there are men

in a whole kingdom!" he wrote musingly. And now microorganisms have been found not only on men's teeth but in every conceivable environmental niche—and also in quite a number that are hardly conceivable at all.

Most living creatures cannot survive extremes of heat or cold. At very high temperatures, the complicated chemicals that are essential for life are destroyed, and at low temperatures the chemical reactions required to provide energy for growth take place too slowly to support life. Nevertheless, some microorganisms are able to live and grow at the freezing point of water (32°F). Particularly spectacular examples are the algae that grow on the surface of Arctic snowfields in such profusion that they color the snow red or green. Furthermore, freezing does not necessarily kill bacteria. Proof of this is shown by the fact that when samples of frozen human waste taken recently from huts used in 1913 by the British Antarctic explorer Sir Ernest Shackleton were thawed, they yielded living intestinal-bacterial cells.

Some bacteria, too, can tolerate extreme

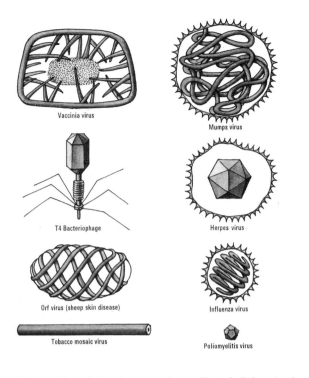

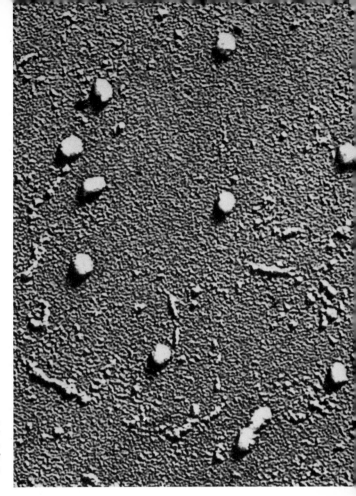

Above: although the viruses are the smallest of all the microbes, these "living chemicals" have distinct, and often remarkably complex, structures. This is what a few of them look like when examined through an electron microscope. Right: an actual photomicrograph of particles of the spherical virus that causes influenza, a common ailment of man.

Vaccinia virus

Mumps virus

T4 Bacteriophage

Herpes virus

Orf virus (sheep skin disease)

Influenza virus

Tobacco mosaic virus

Poliomyelitis virus

heat. The best-known examples live around the boiling springs that can be found in several parts of the world. There are over 10,000 such springs in Yellowstone National Park, Wyoming, where the temperature of the issuing water is nearly 200°F. Yet rod-shaped bacteria are to be found growing and multiplying on the rocks washed by the steamy waters. And the endospores of some species of bacteria can withstand continuous boiling for several minutes, a treatment that would destroy most other forms of life.

In the absence of water, however, even the hardiest microorganism cannot grow and multiply. But many can withstand extreme desiccation. For instance, living bacteria have been grown from spores preserved in dried 17th-century plant specimens. Microorganisms have also been found at extreme depths and heights. Some have been detected flourishing in the sediment at the bottom of the seven-mile-deep Pacific trench, and at the other end of the scale, American spacemen have found fungal spores several miles up in the stratosphere. There are also bacteria that can resist attack by what are to us

poisons and dangerous chemicals. Some species actually produce sulfuric acid from sulfur compounds, and others have been found growing in conditions of extreme alkalinity.

All of this indicates the measure of the extraordinary success of microorganisms as living creatures. They can thrive in some of the unfriendliest environments on earth. Many microorganisms are entirely dependent for their survival on the creatures they live with or in, and they have to be adapted to life in their willing or unwilling hosts in much the same way that bacteria around natural hot springs must be modified to withstand the continual dousing with boiling water.

For some microorganisms, the relationship with the host is wholly destructive. The parasitic bacteriophages, for example, use their bacterial hosts merely as a factory for reproducing themselves; once finished, they leave the gutted hulk in search of new cells to capture. But there are many, many ways in which organisms can live in association with one another, and most of these are beneficial to one or both

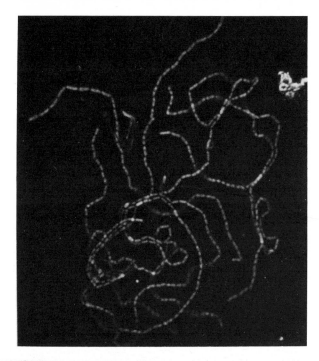

of the partners. We have already mentioned parasitism and saprophytism in relation to fungi, but there is also *symbiosis*: the prolonged, close, and often profitable association of different species. An example is that of the green algae that live within protozoans: the algae gain mobility and shelter, while the protozoans gain whatever excess food the algae manufacture. Another very common form of association is *commensalism*, where, although all the benefits go to one of the partners, the other is not harmed in any way by the relationship.

In the pages that follow we shall see how the five types of microorganism live, and find their nourishment, and we shall try to understand the part they play in ensuring that the elements vital to life on earth are kept continually in circulation. We shall see also how the various kinds of invisible beings have become adapted to living in association, both loose and intimate, with other living creatures.

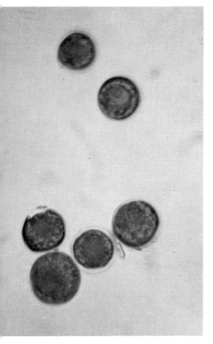

No other living creatures can compete with the microorganisms in their ability to survive in the most inhospitable of environments. They thrive not only in every likely place but also in places where almost no other life could exist. Rocks around Minerva Spring in Yellowstone National Park (left) are continuously washed by steaming water at nearly 200°F; yet rod-shaped bacteria like those at the top of the opposite page grow and multiply there. At the other end of the thermometer, in the freezing heights of the Cascade Range of Oregon, there are so many red-pigmented alga spores (similar to those isolated above in a highly magnified picture) that the snow itself looks red. The red snow shown at right is the result of a profusion of resting spores. The snow turns green when many alga cells are in the growth rather than the resting stage.

Free-Living Microbes

Without the unique abilities and activities of microorganisms, life on earth would soon grind to a halt, for microbes supply vitally needed building materials to living organisms. We live in a world where the essential materials of life are in limited supply; if the materials that provide growth and energy for one generation were to remain imprisoned in their bodies after death, there would be a severe shortage of resources for succeeding generations. Life continues to exist because these materials are recycled. When a plant or animal dies, its remains are pillaged for food by bacteria and other microorganisms, which break down the dead tissue, reduce the materials from which it is constructed to simpler forms, and eventually return them to the environment for further use.

It is chiefly the free-living microorganisms that maintain these natural cycles of growth and decay. Free-living microorganisms are those whose evolutionary path has led them into a life style in which they must find their food in open competition with other forms of life. They do this efficiently because of their versatility and their ability to colonize new environments and to use all sorts of chemicals as food. So now let us consider a few examples of these invisible creatures in search of food.

Protozoans must hunt and catch their food. Thus they need some way to move about; and, as we saw in Chapter 1, they are equipped with a variety of means of movement. Some protozoans—the amoebalike species—both feed and move by means of the same mechanism. They engulf their food as they flow along, surrounding it with protoplasm and incorporating it into the cell. The food may well be a bacterium, which becomes enclosed inside the protozoan together with a small drop of the water in which it has been swimming. Digestive juices are then secreted into this vacuole to break down the food into simpler substances.

The only visible life on this rotting carcass of a horse is a butterfly, which has alighted there for a brilliant moment. But the very word "rotting" implies another kind of life; dead flesh rots because it is alive with microorganisms that break down the tissue and recycle it into the environment.

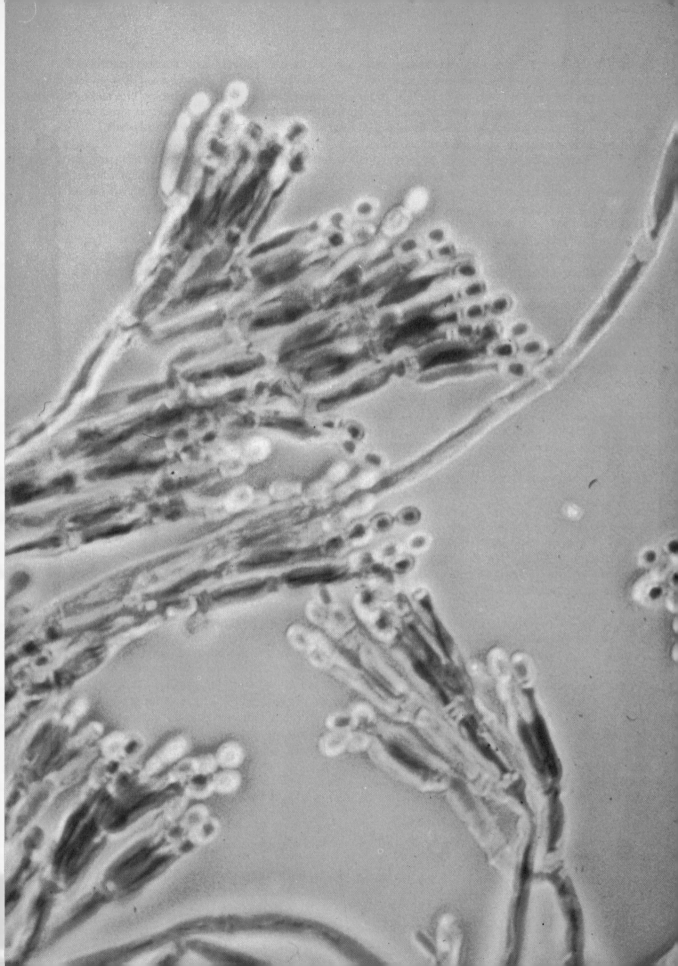

leaf tissues. So yet another group of microbes, which can use these particular acids as food, can now join in the banquet.

While the bacteria have been feeding, protozoans grazing in the film of water on the leaf surface have been consuming the bacteria. Larger soil animals such as earthworms, woodlice, earwigs, and termites help to tear the leaf into smaller pieces, many of which become incorporated in the top layers of the soil. Eventually, all the leaf is decomposed and its substance carried into the topsoil or bound into the bodies of microorganisms, which die in their turn and are decomposed in the soil or eaten by other creatures. And so the life-making substances once locked into the leaf are redistributed. There is a very old saying that "corruption is the mother of vegetation"—proving that long before they discovered the presence of micro-

organisms, farmers recognized that what makes the difference between fertility and infertility is the soil's capacity to release the vital nutrients from dead remains.

One of the most important differences, in fact, between a fertile soil and a mere accumulation of rock particles is the presence in the soil of a thriving microbial population. A fertile soil is either well supplied from the beginning with nutrients in a form that green plants can use, or rich in microorganisms able to unlock the vital nutrients from dead organic remains. Man, of course, makes use of the decomposing abilities of microorganisms in preparing compost—rotted-down heaps of animal and vegetable remains are added to the soil to improve its fertility.

Soil bacteria live chiefly on the surface of the particles that make up the soil, and they adhere tightly to these surfaces. They are present in

Mold, like that on the quince below, is formed by the air spores of free-living saprophytic fungi. One such species, Penicillium *(left), produces penicillin, whose power to kill certain disease-bearing bacteria has proved a boon to medicine.*

very large numbers. Recent measurements suggest that soil from arable land may contain over 100,000 million bacteria in each ounce. Agriculturalists have calculated that if they average this number throughout the top six inches of soil, and if the top six inches weigh about 1000 tons an acre, there may be between 0.6 and 1.5 tons of living bacteria in every acre of good earth. Moreover, representatives of all the other types of microorganism also live in the soil. Some indication of the vast numbers of these simple forms of life is the authoritative calculation that for every 300 million bacteria in good soils there are 3 million fungi and over 60 million other microorganisms, including protozoans and blue-green algae.

One group of soil organisms that we have not yet discussed is the actinomycetes. In some ways these free-living creatures are like bacteria, but in others they resemble the fungi. They form fine, many-branched networks of microscopic threads in the soil, and they produce a substance that scientists term "geosmin," which is responsible for the characteristic musty odor of newly turned earth. A typical example of a soil actinomycete is *Streptomyces*. Perhaps the most striking characteristic of this microbe, and of its near relatives, is the fact that it produces an antibiotic—a substance that kills or prevents the growth of other microorganisms. We shall see in a later chapter how we use this substance, called *streptomycin,* as a weapon against disease.

So far we have seen free-living microorganisms at work chiefly as demolition experts; now let us look at them as creators of the building-bricks of life. Nitrogen, a constituent of protein, is one such building-brick. Although nitrogen is the most abundant gas in the atmosphere, plants and animals cannot incorporate it directly into their protein. Plants can absorb it from the soil only in a watery solution, as a salt of ammonia or as nitrate. Bacteria release ammonia and nitrates into the soil through the decomposition of dead remains, but they also free nitrogen that is imprisoned in plant or animal remains and release it as a gas. Through this process and through *leaching*—whereby rainwater, in draining through the top layers of soil, carries soluble mineral salts down beyond the reach of plant roots—soils are continually being depleted of nitrogen. Soils that support commercial crops also lose nitrogen, of course, because the harvested crop takes a good deal away.

Some nitrogen in the form of nitrates and ammonia is returned to the soil in rainwater—up to five pounds an acre in one year, according to measurements made at Cornell University in New York State. But that is hardly enough to compensate for the amounts lost in other ways. What does make up for the loss is the activities of soil microorganisms that convert gaseous nitrogen into soluble nitrogen-containing substances that can be absorbed by plant roots and turned into protein. Because green plants are the primary producers of food for the majority of living creatures, the importance of such microorganisms cannot be overestimated.

The most important way in which this conversion occurs is through bacteria that live in special parts of the root—the root nodules—of leguminous plants such as peas and beans. But this is in a special kind of association that we shall be discussing in Chapter 3. There are free-living organisms, however, that do fix atmospheric nitrogen. The list of such organisms includes some bacteria and a few yeasts, but the most important contribution to the overall exchange of nitrogen between the living and nonliving worlds results from the activities of certain species of blue-green algae.

In tropical climates, blue-green algae may fix as much as 70 pounds of nitrogen a year for every acre. They abound in the waters of the paddy fields of Southeast Asia, and that fact is of considerable significance to rice growers. In temperate climates, colonies of blue-green algae are found in shallow marine bays where the water is warm, in freshwater lakes where nutrients are freely available, and in the soil. Dark, nearly black patches on the damp surface of the soil in gardens and in flowerpots in greenhouses are almost certain to be the variety of blue-green alga called *Nostoc.*

Because they contain chlorophyll, blue-green algae can manufacture their own food. With this self-sufficiency and with their remarkable resistance to bad conditions, they are often the first form of life to colonize a new habitat. For example, they appeared on the rocks and volcanic ash of the Indonesian island of Krakatoa within three years of the devastating volcanic eruption of 1883, which blew off the entire top of the island and destroyed all its visible life. More recently, between 1963 and 1967, when further violence within the earth's crust threw up the island of Surtsey off the south coast of Iceland, the

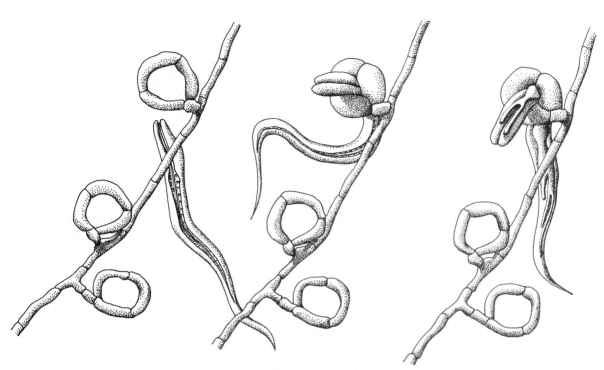

How one type of soil-dwelling fungus snares living prey: the mold forms a sticky network of loops, into which an unwary eelworm blunders; the trap is sprung as the loop's three cells expand and hold the worm in a firm grip; after hours of struggle, the victim dies and is digested by branching filaments of its captor.

barren surface of the new land became the home of blue-green algae within a year.

Nitrogen fixation is by far the most important process through which microorganisms turn a substance that cannot be assimilated by green plants into one that can. But they also help to make other substances available. Sulfur, for example, is a constituent of some proteins essential to living things, but it too can be absorbed by plants only in soluble forms, some of which are not naturally present in soil. Bacteria play a major role in decomposing dead plants and animal remains and releasing soluble sulfur compounds. One of the substances that the bacteria produce in this process is hydrogen sulfide gas—giving rotting materials their distinctive smell of bad eggs.

Almost all the free-living microorganisms that we have discussed so far need to find their food ready-made. Exceptions are the blue-green algae and the true algae. True algae are green plants, and as such are on the other side of the biological fence from microbes, which need to hunt and compete for their food. With the sugar that they manufacture by photosynthesis, together with small amounts of nitrogen, sulfur, and other vital elements, they can undertake the complex chemistry involved in creating the substances required for growth. Once incorporated into green plants, these substances are available to any creature that eats them.

Over half the new plant material created each year is produced by land plants. But the remainder—in specific terms about 50,000 million tons—is produced by water-dwelling green algae.

Green algae live mainly in coastal waters and estuaries—the "nurseries of the sea"—and in inland lakes and ponds. They need water rich in nutrients to survive. Some open-ocean areas also support algal populations, however, especially off the coasts of California and Peru, where winds and currents cause the up-welling of deep ocean water, bringing nutrients to the surface.

Green algae drifting in the sea are called *phytoplankton*—which means plant plankton. The mass of living organisms, plant and animal, that simply float at or near the surface and are carried about by the water's movements are known collectively as the *plankton*. Most phytoplankton live near enough to the surface for sunshine to reach them, so that they can photosynthesize. The phytoplankton have been called the "pastures of the sea," and they are extremely lush pastures. They are eaten by their larger

The Invisible World of a Woodland Floor

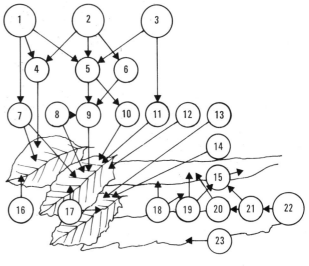

1 Predatory mite
2 Pseudoscorpion
3 Centipede
4 Springtail
5 Psocid (booklouse)
6 Nematode
7 Mite
8 Protozoa
9 Bacteria
10 Algae
11 Earthworm
12 Pillbug (woodlouse)

13 Millipede
14 Sowbug (woodlouse)
15 Fungi
16 Weevil
17 Mite
18 Earthworm
19 Springtail
20 Bacteria
21 Nematode
22 Predatory fungus
23 Earthworm

A host of hungry creatures are involved in decomposing the leaves that fall onto a woodland floor: many of them are invisible microorganisms, others the tiniest of insects, some (such as the earthworms) big enough to be easily visible. But we are seldom aware even of those we can see, because their habitat is hidden in the litter layer or deeper. As indicated by the arrows, various food chains begin with the decaying leaf, which, as it gradually decomposes, sinks down through three layers of soil (whose relative depths are not drawn to scale in this picture): a dark upper layer of humus, which is associated with tree roots; a second, lighter zone that includes both organic material and earth; and the lightest level, composed almost entirely of mineral soil. The system pictured here is based on what happens to beech leaves, but it is fairly typical for any litter breakdowns in woodland areas in the temperate zone.

animal neighbors in the plankton mass—the so-called *zooplankton*—and by other animals, including whales.

Just as vegetation on land changes with the seasons, so the drifting life of the sea in temperate regions alters with the succession of winter, spring, summer, and autumn. In the dark days of winter, there is little light for photosynthesis, and so the phytoplankton, unable to manufacture food for themselves, begin to dwindle in number. The small animals that make up the zooplankton are thus deprived of their natural food, and they too die away. The millions of tiny dead bodies replenish the water with vital nutrients, almost as if the sea were being manured ready for spring.

When spring comes, the bright sunshine stimulates the growth of the algae, and this abundance of food allows the zooplankton to grow and reproduce. Within the space of a few weeks, the planktonic population may have multiplied 10,000 times over. But the warmth that brings on the spring outburst of marine life also has another interesting effect; it heats the surface layers of the water, which become less dense than the lower, colder layers. The plankton in the warm surface water grow and multiply until, having used up all the available nutrients, vast numbers of them die. The dead bodies sink into the cold, lower layers, and so the surviving planktonic population is deprived of the valuable products of their decay.

In this magnified color photograph of diatoms and blue-green algae, the two kinds of organisms are readily distinguishable; naviculoid diatoms are boat-shaped, whereas *Oscillatoria* blue-green algae consist of thin filaments, either in individual strands or matted together into bundles.

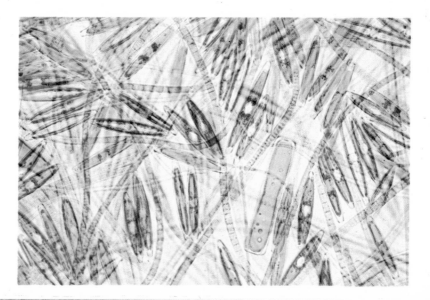

The obvious fertility of the Japanese rice farm below owes much to the micro-organisms that inhabit its watery soil, for the soluble nitrates without which plants cannot survive are produced from gaseous nitrogen by the activities of free-living microbes, particularly the blue-green algae that abound in the waters of Asian paddy fields.

In the autumn, when there is no hot sunshine to heat the water, the surface layers cool off again, and the autumn gales and winds churn the upper and lower layers together. Once more, the temporarily lost nutrients become available to the planktonic community. The tiny creatures respond with an autumnal outburst of life—less spectacular than that of the spring, but enough to enable the plankton to survive the rigors of the winter.

The algae floating in the water are at the mercy of the elements, nourished by whatever nutrients come their way, and are likely at any time to be devoured by the fish and mammals for which the plankton provide a perpetual dinner table. Life is perhaps a little easier for the microbes living on the seabed or at the bottom of a pond. If you look very hard at the bottom of a pond or stream, you are likely to catch a glimpse of the dark brown, fuzzy material that clings to the stones and other surfaces, which is made up of a mixed population of algae, bacteria, and fungi.

Free-living microorganisms, then, have greatly

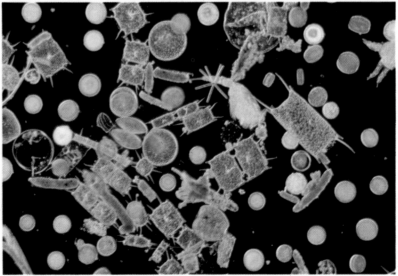

Most of the sea's plant life is found in the phytoplankton, which is composed of minute single-celled plants of many shapes and sizes. Left: various species of diatom, along with animal feces (the round pellets) and one copepod (a tiny crustacean member of the zooplankton). The distinguishing feature of a diatom, whatever its shape, is the hard cell wall that surrounds it like a box. Above: a ship sails through a swarm of Noctiluca scintillans—*dinoflagellates that form pinkish drifts at the surface of calm summer seas. Near right: a close-up of a few of these densely packed creatures, which are luminescent at night when agitated by a passing ship. Far right: another member of the dinoflagellates. Most dinoflagellates have two flagella, extending from the center of the cell.*

influenced the patterns of development of life in many ways. They have also influenced the development of the nonliving physical world. And their influence here is sometimes beneficial, but often harmful, to man.

Certain bacteria, for example, are responsible for the formation of the foul-smelling black mud found on the seabed and in coastal areas, where their life processes create a compound of sulfur and iron that gives the mud its distinctive color and appearance. Apart from its odor, the black mud is comparatively harmless. But the same kind of bacteria whose activities form the mud can also be the corrosive agents that attack iron and steel pipes buried either in the seabed or in the soil. These pipes are open to attack from rusting in any case, but the bacteria help matters along considerably. In the Netherlands, water pipes often have to be laid in land that has been reclaimed from the sea, and they are coated with asphalt to give protection. Nevertheless, the pipes can become so corroded in a year or two that the iron can be cut with a knife as though it were butter. Ordinary rusting without bac-

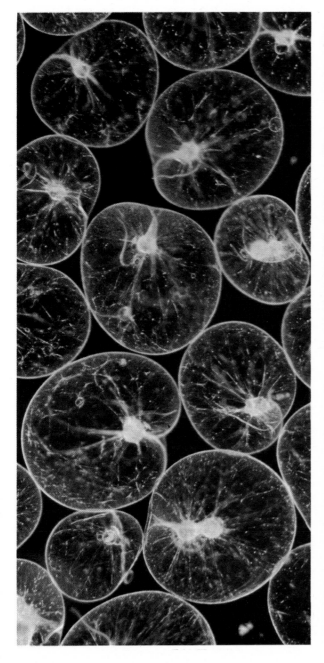

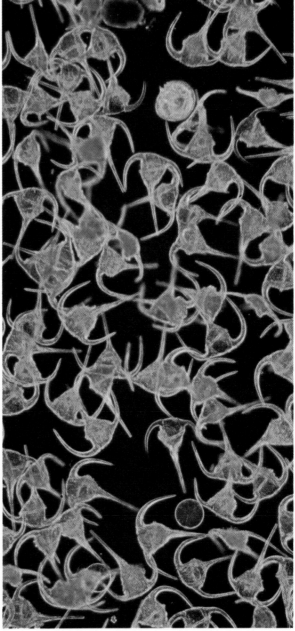

This gigantic snail-like bowl, which lies near the Mexico City Airport, is actually called "the snail" (El Caracol). It gets its fascinating succession of colors from the micro-organisms that live in the salt water that flows through it. The bowl, two miles in diameter, was built by a company that manufactures soda and salt. Brine fed into the spiral flows slowly down the gradient until it reaches the center—a distance of 20 miles in all. From start to finish, the sluggish process takes 6 months, during which time water evaporates and the salt content of the brine becomes more and more concentrated. Meanwhile, the algae and bacteria in the solution change the color of the bowl as its contents get closer to the middle.

terial assistance could not do so swift a job of destruction.

On the other hand, it seems likely that micro-organisms played a partial but significant role in the formation of the fossil fuels—peat, coal, and oil—that we prize so highly. What probably happened, millions of years ago, was that huge amounts of partly decayed vegetable matter became embedded in the mud at the bottom of shallow marshes and lakes. Through the course of time, this vegetation became covered with layers of sand and rock. But bacteria are very efficient at decomposing dead organic remains;

why, then, should this great mass of vegetation have remained undecomposed for the natural geological processes of compression to turn it into fossil fuels?

The answer seems to be that the vegetation was probably waterlogged, and so it is unlikely that there was sufficient oxygen available for bacterial decomposition. The only bacteria that could go to work on the vegetation in that case would have been those that did not require oxygen for life. Such bacteria, we know, produce acids while breaking down organic matter— acids that are lethal to the very bacteria that

Peat, coal, and oil—the compressed vegetation from which we get most of our energy—became fossil fuels partly because of microbial action. Instead of being decomposed, the waterlogged plant life that we now burn was only partly decayed, probably by acid-producing bacteria whose own acid seems to have killed them before they could finish the job. Right: turf cutting in Ireland, where peat is widely used. Although peat—a burnable type of turf— is easier to reach than coal or oil, it is far less energy-productive. Below: offshore drilling for oil in Alaska.

produce them. So decomposition would have ceased at the stage of acid production, leaving the partly decayed vegetation to be compressed and turned into peat, coal, or oil.

It is an interesting fact that the microorganisms that played a role in the formation of coal and oil now play a role in helping man to prospect for these substances. Where there are deposits of coal or oil, a number of gases such at methane, ethane, and propane may seep to the surface and provide nutrients for certain kinds of microorganism that can grow in these gases in the absence of oxygen. Wherever such bacteria seem to be thriving—and they can be discovered through laboratory analysis of the soil—there is a good chance that a fossil-fuel deposit will be found nearby.

The invisible decomposers are still at work forming peat in bogs and marshes. A frequent by-product of this bacterial activity is marsh

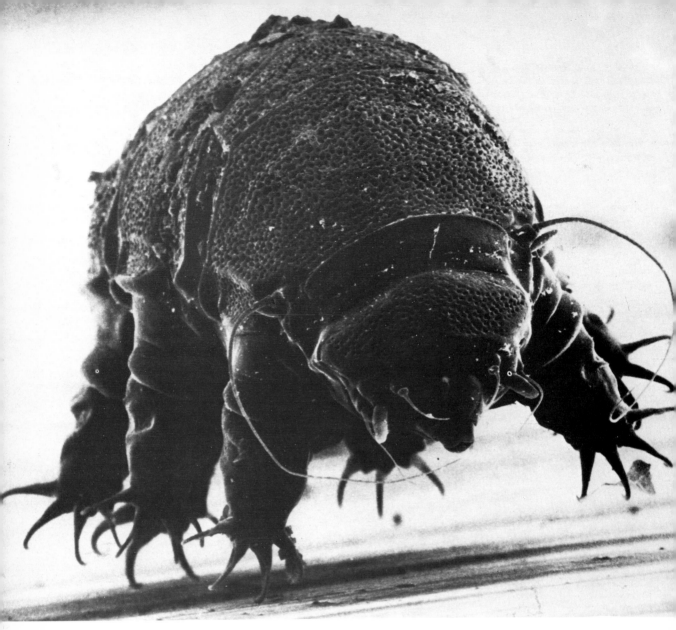

gas, or methane. It can be seen bubbling from peat bogs, where it sometimes ignites spontaneously, giving rise at night to the will-o'-the-wisp that used to be blamed for luring unwary travelers off the safe path and into the deadly marshes. In modern times, methane has proved to be invaluable to mankind, for great pockets of it have been discovered throughout the world, and these are tapped to provide heat and power for domestic use and for industry.

We have concerned ourselves up to this point with free-living microorganisms at work in the soil and at home in the water. What about the atmosphere, the region in the biosphere that we have not yet mentioned? The answer is that for most microorganisms the air is a medium of transmission rather than a habitat. It is full of

Right: shallow standing waters are often clogged with living material (the periphyton) built up chiefly from algae and containing well over 100 different representatives of microorganic plant and animal life. Among them may sometimes be found the strange, primitive monster pictured above. This tardigrade (or "water bear"), though not technically a microbe, is nonetheless too small to be easily seen with the naked eye; a scanning electron micrograph has magnified it to 150 times its actual size. One remarkable characteristic of the tardigrades is their ability to survive very extensive periods of drought. They have been known to return to life after as much as seven years of virtual nonexistence in a shriveled state.

living organisms, but most of them are in their resistant spore form—the well-protected packages that germinate and produce growing organisms only when they alight on a suitably life-supporting surface.

Measurements made not long ago in one of the world's biggest cities, London, indicate that there are as many as 14 living spores in each cubic foot of air outdoors. Indoors, the total may be very much higher, and up to 1000 spores per cubic foot have been recorded. The difference in the numbers of microorganisms in a cubic foot of outdoor air compared with a cubic foot of indoor air may seem remarkable; in fact, however, most of the indoor organisms come from the human respiratory system. Up to 100,000 bacteria may be expelled in a single sneeze, and so it is perhaps remarkable that the indoor figure is not even higher.

Fungal spores are the most common of the microflora in the air, but there are also bacterial cells and spores, virus particles, and occasional algal cells. At some seasons there will certainly be pollen grains from flowering plants, too. It is largely the pollen grains that cause hay fever, but certain fungal spores also have unpleasant effects on people who are allergic to them. Farmer's lung, a complaint sometimes suffered by farm workers, is the result of an allergic response to chemical substances present in the airborne spores of fungi that live on hay.

Fungi in general have developed a wide range

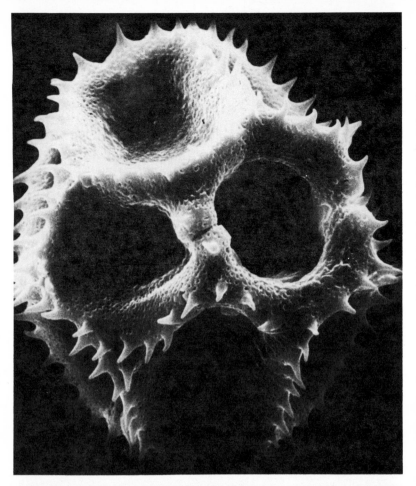

The air is less a habitat for micro-organisms than a medium of transmission, and so its flora consists mainly of pollen grains and fungal and bacterial spores for dispersal and reproduction. This dandelion pollen grain (left), magnified to 3000 times its true size, is a typical example of the kind of airborne pollen that bedevils hay-fever sufferers. but some people are also allergic to fungal spores. Below left: a water droplet lands on the sporophore within the star-shaped outer case of an earthstar fungus, inducing it to expel a cloud of spores for the wind to carry away. A rather more powerful triggering instrument than the drop of water is the combine pictured below. If there is any fungal rust in this American wheat field, spore clouds liberated into the air by the combine may infect other wheat fields hundreds of miles away.

of mechanisms to exploit the air as a means of spore dispersal. Some, such as the mushrooms and toadstools, erect large fleshy structures from which spores are launched into the atmosphere and carried away in the air. Simpler fungi erect only a single threadlike stalk with a ball of spores at the end. In more complex varieties, spores may be ejected like tiny projectiles.

Microbial spores can be carried long distances by wind. Whenever crops are attacked by fungi, there is almost certain to be a vast production of fungal spores. Fungal rust and mildew, in particular, give off great clouds of spores, which may travel a very long way. In America, the fungal disease wheat rust travels north from Mexico and infects crops as far away as the Canadian prairies. In northern Europe, wheat can be infected by rust spores carried by wind from Portugal or Algeria.

Some species of bacteria, if their growth is threatened by the exhaustion of an essential nutrient or by some other sort of harsh condition, form endospores. These can remain dormant for many years, whether in the soil, in water, or in air, but they can germinate to produce normal cells in a matter of minutes. We shall see in Chapter 5 how airborne spores can spread fungal and bacterial diseases.

And so in their quest for survival, in their never-ending battle against the elements and competing organisms for the raw materials of life, free-living microorganisms have changed, and are still changing, the face of the earth. Indeed, they profoundly influence all aspects of life on this planet, and we see their effects everywhere, although they themselves are invisible. In the next few chapters we shall look at the life styles of those microorganisms that are not free-living but that depend on other living creatures for their food and housing.

Plants
and Microbes

Two can live as cheaply as one, they say: and for many plants and animals some sort of close relationship has become a way of life and a hedge against the possibility of extinction. The diversity of nature gives amazing scope for different types of association between living things: animal with animal, plant with plant, animal with plant—and microorganisms with every kind of organism you can think of. Such associations may provide protection against the elements and may vastly improve the partners' chances· of having an adequate food supply.

In this chapter we shall look at associations between microorganisms and plants, ranging from the very loose to the very intimate. Mutually helpful associations of plant and microbe are sometimes helpful to man, for on some of these depends the success of important food crops. And harmful associations can hurt us by damaging our crops, occasionally changing the course of history as they do so.

The fungus that causes potato blight, for example, has had an impact not only on the history of plant disease, but on the history of the world in general. When this fungus ravaged the Irish potato crop in 1845 and in the years following, the consequent widespread famine in Ireland resulted in mass migrations to England and America. It was only later that scientists discovered that the disease was caused by a fungus—a discovery that led to an entirely new understanding of the importance of fungi in plant disease.

But although the question of whether an association is harmful or beneficial to man may be uppermost in the mind of the farmer or the politician (who has to vote the necessary money for controlling the microorganisms that destroy food crops), the ecologist examines each case from a different standpoint. He wants to understand the relationship between different forms

Clover growing in a field. The ability of leguminous plants such as clover to improve soil and secure heavier crops has been known to farmers for 2000 years. We know now that nodules on their roots contain special bacteria capable of turning atmospheric nitrogen into soil-enriching compounds.

Potato Blight

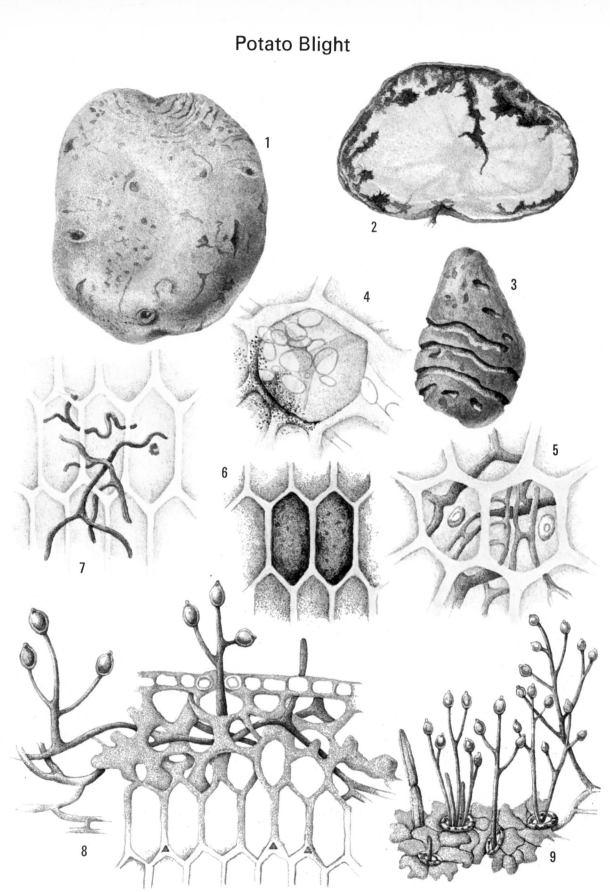

of life, and so he wants to know not only how the microorganism affects the plant, but how the plant affects the microorganism.

Let us start, then, by looking at a remarkable example of two microbial plants, an alga and a fungus, that live together for their mutual benefit. They have become so adapted to their interdependent life style that many people think the result is a single plant. We call this familiar partnership a *lichen*. About 18,000 different species of lichen have been identified, and they show a wide variety of shapes and colors. What you see and recognize as lichen is certainly visible, but we should remember that it is made up of many invisible single-celled components. And the two partners are so tightly bound together that it is impossible to distinguish them as separate organisms without a powerful microscope. Indeed, botanists began to realize that a lichen is a microbial cooperative only a little over a century ago. Before then, when they examined these symbiotic creatures through the microscope, they believed that the green algal cells in the lichen were its reproductive organs.

Typically, a lichen consists of a layer of algal cells embedded in a layer of fungal tissue. Most fungi are composed of a mass of threadlike filaments, and in lichens these filaments are wound tightly together to form the lichen proper, whereas the algal cells lie in a thin layer just below the surface. It is clear that the fungi get their nourishment from the photosynthetic green algae by means of special feeding tubes, which penetrate the algal cells. Although the algae may make up only 5 or 10 per cent of the weight of the entire lichen, they can supply both their own food needs and those of the fungi; in fact, scientists have shown that about half of all the food the alga makes goes to the fungus.

What does the alga get in return for this openhandedness? Some scientists believe that the fungus protects the alga from adverse conditions, especially shortage of water, and also

This illustration shows various stages of potato blight, a fungus related to some of the mildews, and water and bread molds. 1. A potato in the first stages of attack. 2. The fungus spreads into the inner tissues. 3. Fungus growing in a spiral around a small "ash-leaved" potato. In 4, 5, and 7, fungal hyphae can be seen threading through potato cells. 6. The blight has darkened cells in the stem, and threads through leaf tissues (8 and 9) to send up spore shoots through the leaf stomata.

provides a firm base where the alga can grow without much danger of erosion by rain or wind or of an overdose of sunlight, which can kill free-living algae. Others think that the alga gets no benefit at all from the association—in other words, that the fungus lives off the alga without doing it any harm. Still others feel fairly sure that both organisms derive benefits, but they admit they do not yet know what the association does for the alga.

Whatever the nature of the relationship, it is certainly finely balanced and effective, for lichens frequently grow in areas and conditions that few other living things could tolerate (and that includes algae or fungi living on their own). The photosynthetic partner in some lichens is the blue-green alga genus *Nostoc* (which, you may remember, is not a true alga at all but rather akin to a bacterium in structure). Because *Nostoc* can convert atmospheric nitrogen into soluble compounds, a lichen that combines a species of *Nostoc* with a fungus is extraordinarily self-sufficient. It can manufacture its own food, obtain its own nitrogen, resist desiccation, and absorb vital minerals from unyielding surfaces.

This ability to absorb minerals, however, is a hazard as well as a blessing. Lichens are resistant to direct sunlight, drought, heat, and cold, but they cannot tolerate the polluted atmosphere of cities, or any kind of industrial pollution. The apparent reason for this is that the lichens extract minerals and other substances from rainwater; in polluted areas the rain may contain high concentrations of poisonous substances, and the lichens extract these as readily as they absorb useful minerals. Because the lichens have no means of getting rid of the poisons, they are eventually killed off.

In Europe, where lichen growth is used as an indicator of atmospheric pollution, it has been shown that the number of lichen species decreases toward the center of a city. For example, 10 miles from Newcastle-upon-Tyne, an industrial city in northern England, over 50 different species of lichen can be found flourishing; five miles from the city center, fewer than 20 species can tolerate the polluted air; and virtually none are to be found in the center itself. If you see lichen on tree trunks close to a factory, you can justifiably guess that it belongs to one of the very few species resistant to pollutants.

The ability of lichens to absorb and concentrate chemicals from rainwater has had a macabre effect on at least one group of human beings. All of us have a certain amount of radioactivity in our bodies as a result of the testing of atomic weapons in the atmosphere, but the Eskimos have unusually high levels of the dangerous radioactive chemicals cesium 137 and strontium 90 in their bones. The reason is that in the tundra regions rainwater with radioactive fallout has been absorbed by the resident lichens—chiefly reindeer moss and Iceland moss (which are called "moss," but are lichens). These "mosses" are the chief source of food for the reindeer, which in turn are eaten by Eskimos. And so radioactive substances pass through the food chain to the human beings.

Another remarkable thing about the marriage of alga and fungus that forms lichen is its slow growth, along with exceptional durability. Most mature lichens grow only one millimeter or less each year. The age of lichens growing on objects of known age—tombstones, for instance—can be deduced, and it seems they can live for 200 years or more. Naturalists have found Arctic species up to 4500 years old. Indeed, some lichens grow so slowly that geologists use them to estimate the age of glacial deposits. Obviously, for the association to persist for such a long time, the natural balance between the two partners has to be an extremely fine one. Yet it does not take much to upset this balance. It is a severely limited supply of nutrients that apparently causes an alga and a fungus to join forces; if adequately nourished, each of them can survive on its own. Thus, if a lichen is removed from its natural home and given plenty of food and water in a laboratory, the association may break down spontaneously. Either the fungus destroys the alga or the alga overgrows the fungus.

Just as it requires a fine balance for the two microorganisms to sustain their intimate relationship, so the lichen's method of reproduction is a very delicate operation. Some

Lichen (below) is a dual organism, part fungus part alga, both of which are dependent on each other. Often brilliantly colored, as shown on the rocks pictured opposite, lichen are among the first plants to colonize bare inhospitable ground.

lichens—for example, *Xanthoria parietina,* an orange-yellow lichen that flourishes on the rocks by the seashore—have adopted a rather risky technique: the fungus produces dispersal spores, which it shoots out into the atmosphere to be carried away by the wind, and the lichen can be formed again only if the fungal spores germinate alongside cells of the algal partner (in this case a species of *Trebouxia*). If there are plenty of organic nutrients available, any spore can of course produce a healthy fungus on its own; but if nutrients are in poor supply, the fungus will not survive without the help of its algal partner. Other lichens take less risk by producing special dispersal packages consisting of a few algal cells wrapped inside a few fine threads of fungus. These packages are blown away to germinate as new lichens of the same type.

Marriages of the kind that produce lichens are not common in nature. Most microbes that live in association with larger plants enjoy a more casual relationship with their green partners—a relationship that is often so casual, in fact, that its nature can only be guessed at. What is quite clear is that the roots and leaves of green plants usually carry large numbers of microorganisms on their surfaces, and that the soil around the roots is also rich in microorganisms. The area where the roots meet the soil is called the *rhizosphere;* and if you dig up a plant and examine the rhizosphere, you will almost certainly find that its color is slightly different from that of the surrounding earth. This is because microorganisms are at work in the immediate vicinity of the roots.

Scientists have devised ways of looking at plant roots while they are still alive in the soil. Large concentrations of microbes can be seen around the growing root. Soil bacteria evidently cluster in large clumps on the rootlets, even forming fully enveloping sheaths around them, and hordes of bacteria swim in the thin layer of water that surrounds the roots. The fine threads of various species of fungus can be seen winding through the network of roots and rootlets, sometimes putting forth an exploratory thread of living fungus to penetrate the outer tissues of the root. With all this readily available food, it is hardly surprising that predatory protozoans and tiny nematode worms are abundant, feeding on the bacteria and the fungi.

What benefit do the bacteria and fungi draw from their association with plant roots? Chiefly

nutrients, for roots are not watertight and are bound to leak out such substances as sugars, amino acids (the basic components of protein), vitamins, and other substances that microorganisms can use for growth. Some species of *Bacillus* bacteria secrete chemical substances, rather like powerful household detergents, that actually increase the leakiness of plant roots. The large plant, on the other hand, seems to get very little from such loose associations, although it is possible that the presence of the organisms of the rhizosphere in some way helps the plant to absorb chemicals from the soil.

Sometimes the microorganisms around its roots can positively harm a plant. One group of bacteria get their energy through a chemical reaction involving the element manganese. Healthy oat crops need manganese; and if their rhizosphere is overburdened with manganese-using bacteria, the resultant deprivation gives the oats what is known as "gray-speck disease." The cure is to suffuse the soil with a chemical that poisons the bacteria without hurting the oat crop.

The aerial parts of plants—stems and leaves—also have casual microbial associates. The problems that these microorganisms must solve in order to maintain such above-ground relationships are quite different from those in the rhizosphere. The greatest problem is desiccation. It is not surprising, therefore, that most plants with extensive microbial populations on the leaves are tropical in origin. Because of the high humidity of most tropical regions, the aerial parts of tropical plants are often coated with a fine film of moisture, which encourages the growth of microorganisms. A rubber tree, for example, may harbor over 65 million microorganisms on each square inch of its leaf surface. And they are well fed, too. The waxy cuticle on tropical plants is thinner than that on temperate species, for water loss is much less of a problem, and so various leaf constituents ooze out onto the surface and become available to the trespassers. Flowers that produce sugary nectar to attract the insects for pollination are also ideal sources of food for microorganisms.

The invisible creatures that live on leaves and stems do little harm, even though some of the bacteria produce gums or slime to help themselves adhere to the leaves during heavy rainfall. But there is at least one exception to this general rule. When aphids attack a rosebush,

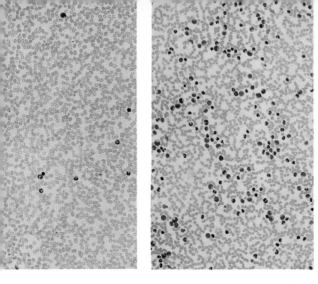

Photomicrographs of blood taken from a leukemia—blood cancer—sufferer (above right) compared with normal blood (above left). Scientists think that strontium 90 might be a possible cause of the disease. Reindeer moss (right), which absorbs and concentrates minerals from rainwater, contains a high proportion of strontium 90. The moss, which is in fact a lichen despite its name, is eaten by reindeer (below), a principal, and probably dangerous, food source of the Eskimos.

How Microbes and Plants can Benefit Each Other

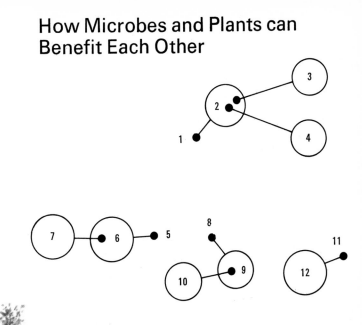

Lichen – alga/fungus association
1 Lichen growing on bark of birch tree.
2 Enlarged cross section through lichen.
3 Algal cells.
4 Fungal hyphae.

A lichen is a symbiotic association between an alga and a fungus. The algal cells are the photosynthetic components of the lichen. These are embedded in a mass of fungal tissue, which obtains its food from the algal cells and, in return, provides protection for the thin algal layer.

Ectotrophic (external) mycorrhiza – root/fungus association
5 Hyphae of fly agaric fungus surrounding roots of birch tree.
6 Enlarged portion of mycorrhizal birch tree roots.
7 Cross section through mycorrhizal roots showing the fungal sheath as a dark outer band.

A mycorrhiza is a symbiotic association between the roots of a higher green plant and a fungus. An ectotrophic mycorrhiza is one in which the fungus forms a sheath around the outside of the root. The fungus gets all the sugar it needs from its host and the green plant obtains mineral salts from the fungus.

Root nodule of legume – root/bacterium association
8 Root system of clover showing nodules, which contain bacteria.
9 Enlarged view of clover root system.
10 *Rhizobium* bacteria from nodules of clover root.

Root nodules are formed as a result of a specific symbiotic association between a leguminous plant and bacteria of the genus *Rhizobium.* The bacteria collaborate with the plant in the process of converting gaseous nitrogen into nitrogen compounds that both the plant and the bacteria need for nourishment.

Endotrophic (internal) mycorrhiza – root/fungus association
11 Root system of orchid.
12 Longitudinal section through orchid mycorrhiza showing fungal hyphae within root cells.

An endotrophic mycorrhiza is one in which the fungus actually penetrates the tissues of the root. The relationship between the two partners is not fully known but it seems that the green plants involved do need the fungi for successful growth.

63

they pierce the leaves with their sharp mouth-parts in order to get at the sap. As they feed, they deposit droplets of sticky honeydew on the leaf surface. These droplets are rapidly colonized by fungi, and the growth of these fungi, together with the formation of fungal reproductive spores, forms a dense, sooty-black coating on the leaf surface. This undermines the capacity of the rosebush to manufacture its food, for the fungal coating not only reduces the amount of light that reaches the photosynthetic tissue of the leaf but also blocks the stomata—the pores through which carbon dioxide reaches these tissues.

Even this last relationship is a highly casual one, however, for although the rosebush is a host to the fungus, it is the honeydew, not the plant, that nurtures the microorganisms.

Now let us look at examples of much more specific relationships between green plants and microorganisms—associations, first of all, that truly benefit each partner in the arrange-ment. The most important such association—to man, at any rate—is found in peas, beans, and other leguminous plants that have formed a remarkable alliance with bacteria of the genus *Rhizobium*. So intimate and specific is this relationship that the plant forms special lumpy structures, or nodules, on its roots to house its bacterial partner. You can see these clearly if you pull up a legume and wash the roots.

Such a sophisticated arrangement must be based on peculiarly valuable advantages to be had from the association. The key to the matter is the fact that the *Rhizobium* bacteria col-laborate with the plant in the act of converting gaseous nitrogen into nitrogen compounds (a process called *nitrogen fixation*) that both the plant and the microorganism need for nourish-ment. Unlike certain free-living nitrogen-fixing bacteria or the blue-green nitrogen-fixing alga that forms part of several types of lichen, neither the *Rhizobium* bacteria nor the leguminous

Below: runner bean plant awaiting harvesting. They belong to an important group of plants called legumes. Below right: swellings, or nodules, which are packed with bacteria, grow on the roots. The bacteria are fed by the plant and in return they turn atmospheric nitrogen into compounds that can be used by the plant.

plants in whose root nodules they live can fix nitrogen by themselves. They can do it only in close partnership with each other.

Actually, both the bacterium and the plant can survive without each other if each is supplied with its own source of absorbable nitrogen, and so the relationship is in no way obligatory. (In fact, if we did not know that the root nodules fix nitrogen, we might easily assume that the bacterium is harmful to the plant, because the nodules look like abnormal tumors.) But although the two organisms are not absolutely necessary to each other, they do much better together than apart. In nitrogen-deficient soils, for example, legumes with root nodules grow far faster than those without the nodules.

The inside of the mature nitrogen-fixing nodule in which the bacteria live is bright red in color because it contains a protein that is like hemoglobin, the iron-containing protein that gives red blood its color. Neither the plant nor the microorganism alone is responsible for the formation of this red pigment; it is formed as a result of cooperation.

Exactly how the plant and the bacteria work together to fix nitrogen remains a mystery in spite of extensive scientific study, nor is it known why peas and beans should be so especially favored. Some scientists point out, however, that many leguminous plants are tropical in origin, and their ancestral home was the tropical rain forest, where the soils had been washed virtually free of nutrients by the ceaseless rainfall. Thus, they would have had to find a different way to acquire such nutrients.

Although we do not yet know how they work together, we do know how the plant gets its colony of bacteria and forms its root nodules. Like those of all land plants, the roots of leguminous species allow some nutrients to pass into the surrounding soil, and this stimulates the growth of the bacteria in the rhizosphere. A special layer gradually develops around the outside of the leguminous root, thus preventing the nutrients from draining away into the soil. As the layer develops, the rhizosphere bacteria become concentrated in an ever-growing band. Within this enclosed space, the growth of bacteria of the *Rhizobium* type is favored, but the other microorganisms die off. The *Rhizobium* bacteria grow and multiply, infecting the root through the root hairs (hairlike projections

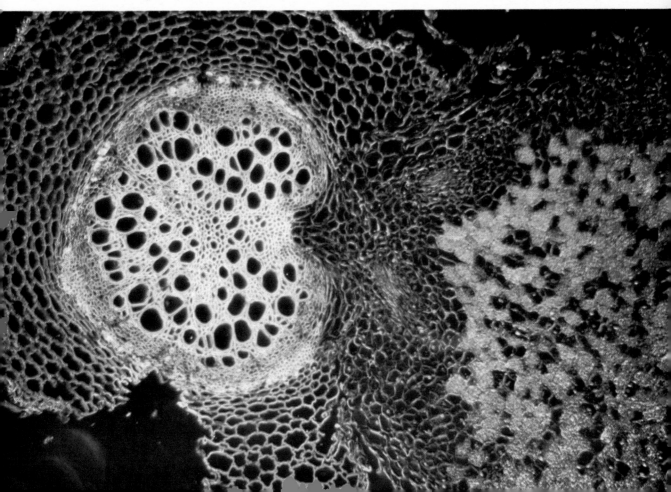

from the root surface, which play an important part in water and mineral salt absorption).

Some of the normally rod-shaped bacteria become transformed into spherical "swarmer cells" covered with whiplike flagellae for swimming; and they wriggle their way through the layer surrounding the root, and so enter the plant's root hairs. What seems to happen is that *Rhizobium* produces a chemical substance, probably a hormone, that affects the root hairs of leguminous plants in such a way as to "soften" them for the invasion.

Once inside a root hair, the swarmer cells proliferate and grow, forming a so-called "infection thread," which leads down the root hair to the body of its cell. There are always a number of especially receptive cells in the roots of these plants, and if one of these is infected by the swarmer cells, it begins to divide and form a nodule. Eventually, any such nodule becomes filled with a new form of *Rhizobium,* known as *bacteroids,* which are swollen and often branched.

Leguminous root nodules confer three separate sets of benefits—first, on the plant, which gains a steady source of nitrogen in a readily usable form; secondly, on the bacteria, which gain shelter and an easy supply of sugars; thirdly, on humanity. Long before he understood the biology and biochemistry of nitrogen fixation, man was making use of the nitrogen-fixing properties of leguminous crops to maintain the fertility of his fields. The ability of plants such as vetch and lucerne (or alfalfa) to enrich soil was known in biblical times. It formed the basis of the crop-rotation schemes developed during the Middle Ages, when farmers learned to grow clover and other legumes and to plow them back into the soil before sowing the main crop.

Nodules fix very large amounts of nitrogen. A healthy lucerne nodule can more than double its own nitrogen content in a day. Worldwide, about 1000 million tons of nitrogen are fixed annually by living organisms, and most of this comes from leguminous plants. New Zealanders, for example, rarely add nitrogen fertilizers to their soil. Instead, they grow a lot of clover, and it has been estimated that the bacteria-filled nodules in such clover-planted soil fix as much as 500 pounds per acre of atmospheric nitrogen every year. Even though the equivalent figure for nitrogen fixation by root nodules in most areas is only a little more than half as much, it is still true that the leguminous plant-

Rhizobium partnership ordinarily produces an annual amount of useful nitrogen compounds equal to a field treatment of three quarters of a ton of the best nitrogen fertilizer.

A few other types of plant also have root nodules containing nitrogen-fixing bacteria. Among them are the sea buckthorn and the bog myrtle. Some tropical plants have *leaf* nodules, which contain nitrogen-fixing bacteria of a different genus from *Rhizobium*. There are also some similar, but less clearly defined, nitrogen-fixing associations, such as that between one kind of water fern and the blue-green alga *Anabaena*. The leaves of the fern have small mucilage-filled chambers that contain these microscopic algae, but the relationship is less obviously a mutually dependent one than the legume–*Rhizobium* association. *Anabaena* seems

Below right: a grain of soil (magnified 640 times) containing some of the vast population of soil microorganisms which, by breaking down dead plants and animals, release substances essential to plant life. After intensive crop-growing, the soil microorganisms need time to build up fresh supplies of nutrients. Early farmers knew only that by periodically resting their fields they could improve crop yields. Top right: the three strip fields in this plan of a typical European feudal village were worked in yearly rotation—the strips worked by one villager are marked in red. The fields are colored to indicate plowing sequence: yellow and brown are spring and fall plowing, pale yellow is fallow, or resting. Below: medieval peasant farmers in their fields.

able to fix nitrogen equally well whether inside the host plant or outside. Then, too, there is a nitrogen-fixing bacterium commonly found on the leaves of tropical plants, but we do not yet know how interdependent the two organisms are. It may be that this bacterium represents a halfway stage between the free-living style of life and the intimate, specialized partnership.

So far we have concentrated chiefly on the associations between bacteria and plant roots. Fungi and roots also live together, and their associations are no less binding and significant. The name for a symbiotic association of fungi and the roots of large green plants is *mycorrhiza* (which means literally "root fungus"). It seems likely that the roots of a high proportion of land plants share their subterranean lives with one fungus or another. In a forest, for example, almost every tree will have a fungal partner wrapped around its roots. In forests of conifer, beech, or oak trees, the fungus forms a sheath around the roots. Much more common, however, is the kind of mycorrhiza in which the fungus actually penetrates the roots, as happens with

orchids. Yet we know a good deal more about the relationship between the forest trees and their external fungal associates than we do about the more common ingrowing root fungi.

Three soil conditions must be met before mycorrhizal fungi will develop readily on tree roots: the soil must be well-ventilated, contain an abundance of decaying plant matter, and be comparatively poor in inorganic plant nutrients. These conditions are relatively common in soils that support plant life, and it is not surprising that so many plants have their associated mycorrhizal fungi. The benefits to the partners are obvious: the fungus gets from its host all the sugar it needs for growth, and the tree gets a greater share of whatever mineral salts are to be had from the soil, for fungi are very good at accumulating such substances.

In soils that are poorly supplied with inorganic nutrients, trees with mycorrhiza thrive, whereas those that are uninfected do not. The mutual dependence between the two partners is so complete that in forest nurseries the appropriate fungus is often introduced into the soil arti-

Mycorrhiza is the fungal equivalent of the nodules that form on the roots of leguminous plants. The picture above shows the roots of the coral root orchid covered with the commonest type of root fungus, endotrophic mycorrhiza, which penetrates into the cells of plant roots. The fungus is important to many plants and especially the heaths, such as those seen on the right, growing in Glen Spean, Scotland.

ficially if it is not already there. The difference between the trees with mycorrhiza and those without can be dramatic. If Virginia pine-tree seedlings grown on Iowa prairie soil are inoculated with mycorrhizal fungi, they can build up twice as much living substance as do similar seedlings with uninfected roots.

One rather interesting fact about the different functions of the two symbiotic associations of root nodules and mycorrhiza has been observed in the forests around Yangambi in Zaïre. The roots of leguminous trees in the very thick Yangambi rain forest are mycorrhizal but have no root nodules; but in the sparser vegetation of an adjacent open forest, the roots are nodulated but not mycorrhizal. The explanation seems to be that in the intense competition for nutrients from the rain-forest floor only mycorrhizas are successful, whereas competition in the open forest is chiefly for nitrogen, which the trees obtain through bacteria-infested root nodules.

Unless you uproot a tree, of course, you cannot tell whether its roots are infected with mycorrhizal fungi. All you can see in any case

is a thickening and branching of the young rootlets. In the autumn, however, the telltale sign of some types of mycorrhiza is the appearance near a tree of toadstools.

Certain types of fungi, incidentally, are specific to particular types of tree. For example, the very poisonous fly agaric, with its white-spotted bright red cap, often appears in stands of silver birch, thus betraying the fact that the fungus is in league with the birches' roots. Other genera of fungi that form mycorrhizal associations with trees include the fleshy, often brightly colored *Boletus,* and *Lactarius,* which puts forth a toadstool that yields a white milky substance when its skin is broken.

Although we know less about the kind of mycorrhiza in which the fungus penetrates the roots, it seems clear that heath, heather, bilberry, and other moorland plants need these penetrating fungi for successful growth. Heaths and heathers introduced into a garden where none have grown before may fail if the particular species of fungus they require is not present in the soil. Instead of staying in the root, these

fungi spread throughout the plant, and may even get into the seed of some berries (bilberries, for example), thus ensuring that the fungus will be there when the young plant begins to grow. But this type of mycorrhiza is not always beneficial. Some ecologists, in fact, consider it to be a form of controlled parasitism rather than of symbiosis.

The relationships that we have been examining are, if not always mutually advantageous, at least harmful to neither plant nor microorganism. Now, though, let us consider some examples of associations in which all the profit goes to one party in the arrangement, and the other gets a very bad bargain, often being totally destroyed in the process. The gainer is always the microorganism, the loser always the plant, which becomes diseased as a result of the association. Plant diseases are important in nature and in commerce. In a recent year, the United States Department of Agriculture estimated that diseases of crop plants—most of which could be attributed to fungi and bacteria— caused losses of about $3250 million in America. It has been authoritatively suggested that microbe-induced plant disease costs the world's farmers about 12 per cent of their annual crop.

Although it is chiefly fungi and, to a lesser extent, bacteria that are responsible for this devastation, viruses play an important role.

Many viruses are known by the name of the plant they attack and the kind of disease they cause. Among the well-known plant viruses are the tobacco mosaic, potato X, alfalfa mosaic, tomato bushy stunt, tobacco necrosis, tobacco ring spot, and turnip yellow mosaic. Well over 300 such viruses are known to attack and damage crop plants. Plants infected with a virus commonly grow more slowly than uninfected ones, and the tissues of the stem, roots, flowers, fruits, or leaves usually show distinctive changes. They may change color, and the leaves may become spotty, streaky, or stained with rings of light green, yellow, white, brown, or black; hence the familiar names "mosaic," "mottle," "streak," or "ring-spot" virus.

There is no recovery from virus disease. Although a plant may not be killed outright, it may be so weakened as to be of little use in agriculture and unable to compete with healthy plants in nature. Exactly how viruses infect their hosts and what they do once inside, we shall learn later on when we discuss their associations with animals, for the processes are very similar.

The effects of viral infection are not always altogether bad. In tulips, for example, such infection may cause a variegation in the color of the blooms that enhances their beauty and is therefore much prized. Some varieties of tulip have been maintained for generations in the virus-infected state, with the infection transferred from one generation to the next. But such happy results of viral infection are rare: The more common result in most plants is sickness, possibly followed by death.

Tobacco plants are usually infected with tobacco mosaic virus accidentally by field workers, who carry the infectious particles from plant to plant on their hands. Viruses that attack other crops may be transmitted by animals or through infected soil. The same virus may infect several different kinds of plants, causing different disease symptoms in each, and one virus may have many different strains or subtypes, each producing symptoms of differing severity in plants of the same type.

The bacteria that cause disease in plants are also usually able to attack a wide range of plants and damage each in a different way. Common bacterial diseases include crown gall, which occurs on fruit trees, fire blight, which damages fruit trees (especially pear trees) and ornamental trees, and angular leaf spot, which can kill tobacco plants. All these and many more ailments result from bacterial infection. The fire-blight bacteria are carried from tree to tree by pollinating insects. The bacteria that cause angular leaf spot enter tobacco leaves by means of the stomatal pores, and each bacterial colony then feeds upon, and destroys, a small angular piece of leaf. The bacteria, like the viruses, get their sustenance from the host plant; the disease from which the plant suffers is simply the process of destruction that gives life and energy to the parasitical microorganisms.

The most destructive diseases in plants are caused by fungi, whereas, conversely, the fungi are relatively unimportant in their associations with man and other mammals, from whom it is the viruses and bacteria that exact the highest toll. The associations between plants and fungi that result in disease are either loose relationships, which permit a particular fungus to attack a wide range of hosts, or specific associations, in which a particular fungus owes its existence to its ability to live in or on a single type of host plant and no other.

Fungi attack the roots, leaves, stems, branches, flowers, and ripe fruits of plants. Among the fungi that feed on the roots of plants are those that cause various kinds of potato disease; some that are responsible for a serious disease of pine-tree plantations called butt-rot; and an extremely destructive type that eats into the seedlings of cereal crops. When this latter fungus attacks wheat, it brings on the commercially catastrophic disease known by the appropriate name of "take-all." Fruit trees, too, are extremely susceptible to fungus disease. If you notice that the leaves of your plum trees have a curious silvery look, the tree is infected by *Stereum purpureum,* a fungus that produces purple reproductive structures on the branches that it destroys. Apple trees are affected by a less colorful fungus, which makes its home on the bark and forms a canker that completely encircles the branch.

Some of the fungi responsible for the most severe diseases of crop plants have complex and remarkable life cycles. *Puccinia graminis,* which causes black rust (or stem rust) of wheat, is one such fungus. It produces no less than five different kinds of reproductive spores during its life cycle and depends on a specific and intimate relationship with two totally different kinds of plants for its survival. Let us look at its life cycle in detail.

As black rust, *Puccinia* lives on wheat, but it can get there only via the common barberry. The barberry plant is infected by dispersal spores in the spring; these, the first of the five different kinds of spore, are called *basidiospores.* The airborne basidiospores land on barberry leaves, and germinate in tiny drops of water on the leaf surface by putting out germination tubes, which form a suckerlike structure on the leaf surface. While the edges of the sucker clamp fast to the leaf, the center portion grows until it punctures the surface and gains entrance. Once inside, the fungus produces a branched network of fungal threads, which feed by thrusting fine tubes into the surrounding living cells. Attack by a single basidiospore leads to a local infection only a few millimeters across, and in this spot the leaf cells are not killed; instead, stimulated to new activity, they produce a swollen region. Tiny flask-shaped structures, now appearing on its upper surface, contain a new kind of spore, which is called a *pycniospore.*

The pycniospores are coated with a sugary substance that attracts insects. Then, when an insect carries them from one infected area of the leaf to another, a fertilization process takes place and the fertilized cell develops into a spore-producing structure on the underside of the leaf, which makes still another type of spore: the *aeciospore.* (This exchange of spores is a primitive kind of sexual reproduction, and it is interesting that *Puccinia* should have evolved a mechanism for pollination similar to that developed by flowering plants.) The aeciospores are liberated into the air. They cannot survive on barberry plants even if they land on them. Where they can and do flourish, however, is on wheat.

An aeciospore that lands on a wheat plant germinates in a drop of water on the leaf or stem surface, putting out fine germination tubes that enter the plant through the stomata. The thread-like fungus ramifies through the plant and eventually develops yet another spore-producing structure, which ruptures the outer layer of the wheat leaf or stem in order to disperse its *uredospores.* These are reddish brown and give the diseased plant its characteristic rusted appearance. Uredospores are dispersal spores; like the aeciospores, they can be airborne for long distances without losing their ability to germinate. They can infect only wheat, not barberry. And so the disease is spread from one wheat field to another.

Toward the end of the season, the production of uredospores gives way to the formation of dark-brown *teliospores,* which are resting spores. They remain attached to the wheat stalk and are inactive over the winter. The following spring, they germinate to produce basidiospores again. These thin-walled, short-lived spores can germinate only on barberry leaves, where they start the cycle once more.

Wheat rust is chiefly a problem in the wheat fields of North America. The plant is not killed by the fungus, but its ability to manufacture food is greatly reduced, and it may wilt and shrivel. When harvested, the diseased wheat may throw up great reddish-brown clouds. It is sometimes possible to eliminate the disease by destroying barberry plants in areas around the wheat fields. But even in Canada, where barberry is rare, uredospores may be blown from parts of the United States hundreds of miles away.

Not only can fungus diseases devastate food crops, but they may have unexpected side effects on the people who eat contaminated food. The village of Pont St. Esprit, France, became notorious in 1951 after a number of its inhabitants were struck down by a mysterious illness, characterized in many cases by madness and eventual death. The illness was later diagnosed as ergotism, a poisoning caused by bread made from rye flour infected with ergots, which are the spore-producing structures of a fungus called *Claviceps purpurea*.

It is rarely possible to cure plants that have been stricken with fungal or viral diseases. Prevention, therefore, is the watchword among knowledgeable farmers, who carefully burn dead and diseased branches and plow infected leaves into the ground. Fungicides are useful in controlling a number of such seedborne diseases as loose smut, a disease of oats. But perhaps the most sophisticated technique is use of plants that are inhospitable to the microorganisms.

Some species of barberry, for example, resist colonization by the rust fungus because their leaf surfaces are too thick for the fungal germination tube to penetrate. Furthermore, some plants produce chemical substances that kill the germinating fungus as it lies in its drop of water on the leaf surface. An interesting example of one type of chemical defense is found among the flaxes, some varieties of which can suffer from wilt as the result of a fungal association. The resistant species of flax fight off the invader by producing prussic acid from their roots (albeit in tiny quantities). This not only helps to stop the specific dangerous microorganism from growing in the rhizosphere, but also stimulates another kind of soil fungus to produce certain substances that also destroy the dangerous kind.

Now we have seen something of the ways in which microorganisms live with and on green plants, from the very simple, loose associations that exist in the rhizosphere to the extremely complex life cycle of the rust fungus. We have seen that the organisms of the invisible world can produce highly dramatic effects, often causing their partners to produce specific chemical substances either for defense or for the benefit of both.

In the preceding chapter we considered the influence of free-living microorganisms on the great natural cycles of the earth. Now that we also know how microorganisms are involved in the part played on earth by plant life, our next step is to examine their effect on, and associations with, animals—starting with small animals.

Top: a tobacco leaf infected with the tobacco mosaic virus. The spots are areas of dead cells. Above: electron micrograph of tobacco mosaic virus. Right: the virus is spread accidentally by tobacco plantation workers who carry the invisible infective particles from plant to plant on their hands.

Barberry

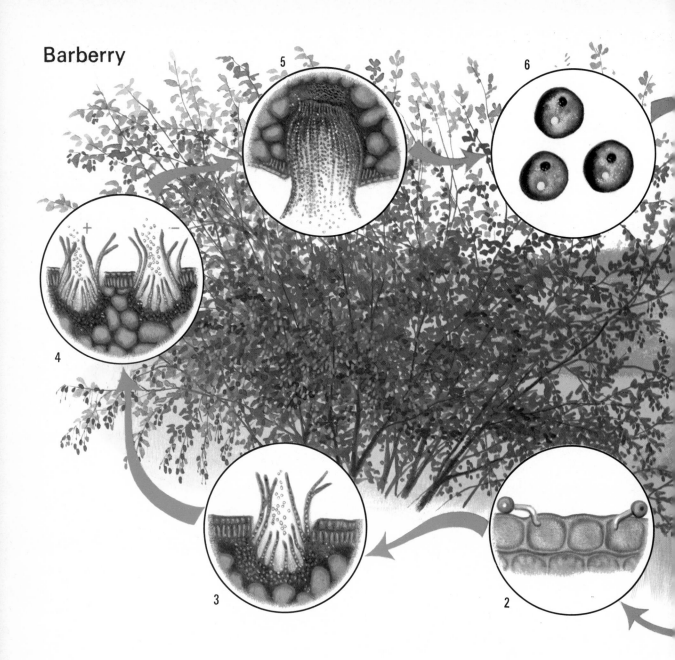

The Life Cycle of Black Rust
(Puccinia graminis)

1 Dispersal spores called *basidiospores* are liberated into the air in spring.
2 Basidiospores germinate on barberry leaves and grow inside leaf cells.
3 Fungal cells multiply to produce structures in which *pycniospores* develop.
4 Pycniospores carried by insects from one infected site to another undergo a process of fertilization.
5 The fertilized cell develops into a structure on the underside of the barberry leaf that produces *aeciospores*

6 The aeciospores are liberated.
7 Aeciospores germinate on wheat stems or leaves and grow through stomata.
8 Structures producing *uredospores* appear on the wheat stem or leaf. These airborne spores infect other wheat plants during the summer.
9 In late summer *teliospores* are produced. These resting spores remain inactive during the winter attached to the wheat stem.
10 The teliospores germinate in the spring to produce basidiospores.

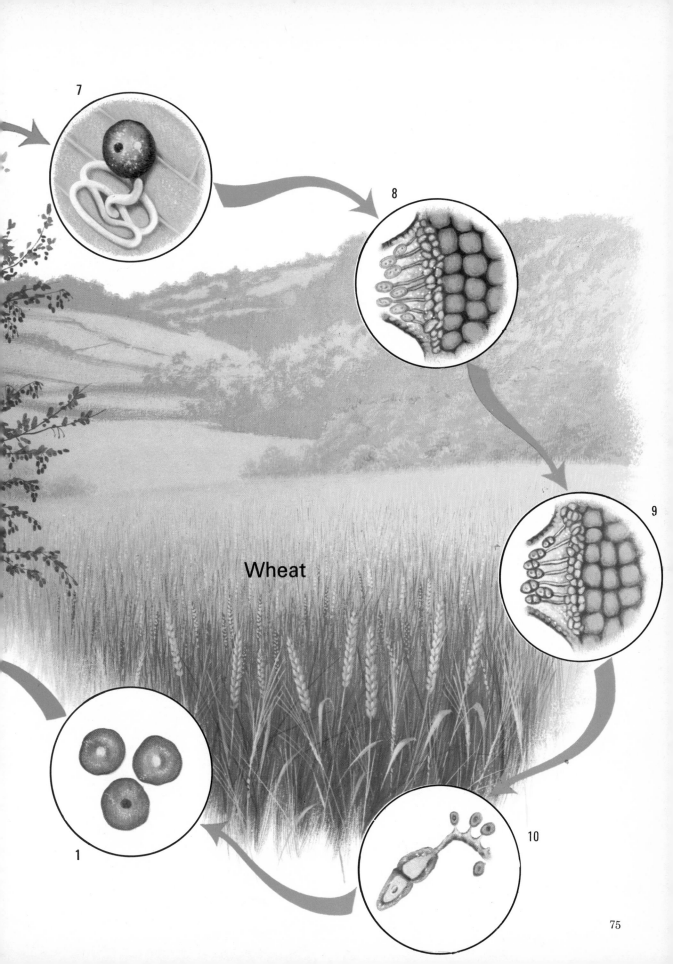

7

8

9

Wheat

1

10

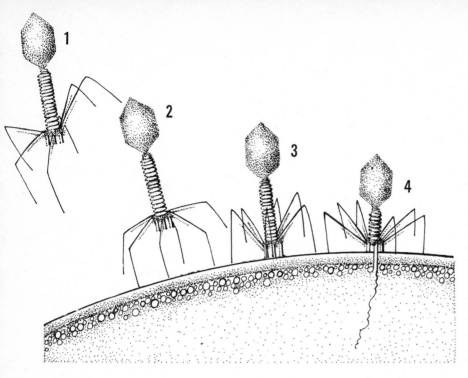

The creature looking like a moon-landing vehicle in the drawings, left, is a bacteriophage, a virus parasitic on bacteria. Stage 1 shows the bacteriophage approaching a bacterium. In stage 2 the tail fibers have made contact with the cell surface. In stage 3, the tail fibers lower the tubular "tail" of the bacteriophage, bringing it into contact with the cell wall. In stage 4, the tail pierces the bacterium's wall and, like a living hypodermic syringe, injects the material from its "head" into the body of the bacterium. The bacteriophage's head contains the hereditary material that will take over the bacterium and direct it to make new phages. Right: electron micrographs of an actual bacteriophage (magnified 650,000 times) illustrate stages 1 and 4 of the diagram at the left.

gut is parasitized by just such a bacteriophage, which has been code-named T4, and of which scientists have made a detailed study. The head of T4 contains the hereditary material—all the information it needs in order to commandeer the living machinery of the bacterium and divert it toward the construction of new phages rather than new bacteria. It was once thought that the phages make their way into the host bacteria by swimming with their tails, approaching the bacterial cell head-first, and chewing their way in. In fact, however, the T4 phage looks rather like a moon-landing vehicle, and recent research has shown that it alights on the surface of its victim like a moon-landing vehicle touching down: tail-first, with the tail fibers attaching themselves to the surface of the bacterium.

Once it has landed, T4 becomes a living hypodermic syringe, and injects its hereditary material into the bacterial cell. The anchored fibers begin to contract, thus bringing the phage's hollow tail into contact with the cell wall. Then in some way that is not understood, but that probably involves the production of a chemical substance, the tail pierces the bacterial wall, and the hereditary material contained in the head is injected directly into the bacterial cell. The empty shell of the virus now falls away and is lost.

There is a big difference between all other kinds of microbial association and the relation-

ship of a bacteriophage to a bacterium. The hereditary material of the phage is the only part of it that invades the host cell, unlike other parasitical associations in which whole micro-organisms enter and live in or on their hosts' cells or bodies. There is, after all, no need for the entire organism to enter the host cell, for the life of the phage within the host cell is bound up in what happens between the phage's hereditary material and that of the host.

The phage's hereditary material can behave in either of two ways once it is inside the host cell. If it is a *virulent* phage, it sets to work immediately to direct the machinery of the host cell toward the production of new phages. Exactly how it happens remains obscure, but new phages soon appear in the host cell; and after several hundred have been formed, they are released as the host cell bursts and dies. The bacterium does not burst merely because it becomes overcrowded with phages. Apparently, the phages produce a chemical that breaks down the cell wall from within, just as a chemical in the tail of an infecting phage pierces the wall from outside. Thus the virulent phage invades a bacterium, where it multiplies many hundred-fold, and its progeny kill their host before moving on to infect other bacteria.

The so-called *temperate* phages may behave like virulent phages and destroy the cell they invade, or else their hereditary material may behave

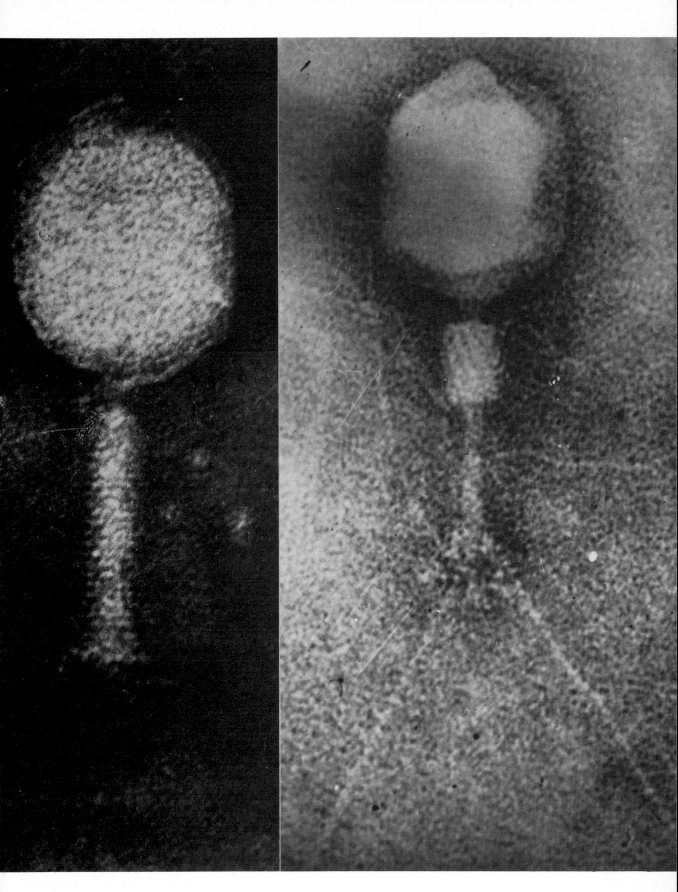

quite differently by becoming combined with the hereditary material of the host. In this latter case, the bacterium continues to live; but when it reproduces itself, it reproduces the phage's hereditary material as well as its own. The particular type of behavior—virulent or temperate—that such phages display depends on the host. Some bacteria evidently permit the combination of their hereditary materials, whereas others do not. If a temperate phage invades a bacterium that is not susceptible to the combined heredities process, the phage destroys the cell while reproducing only itself.

There are certainly few more remarkable associations between organisms than the one in which genetic materials of a phage and its bacterial host are linked together. It is as strange as if a parasitic flea were able to inject its genetic material into a cat, so that the cat's kittens would then have the hereditary characteristics of both animals. And an interesting sidelight is that the temperate phages can add hereditary material not only to bacterial cells, but to animal cells as well. It is thus possible that we may someday learn how to use them to repair genetic defects in higher animals, including man.

For example, one of the numerous diseases caused by faults in the genes of some human beings is galactosemia, a potentially fatal disease that can cause infants to be dwarfed, malnourished, and mentally retarded. It is caused by failure of an individual's cells—or, rather, the hereditary material within the cells—to manufacture a certain chemical necessary for the digestion of certain sugars. Remarkably enough, one bacteriophage, code-named *lambda,* has the ability to produce this very chemical.

In a notable experiment carried out in 1971, when scientists of the National Institutes of Health and Mental Health in Bethesda, Maryland, introduced phage lambda into cells taken from a galactosemia sufferer, the infected cells seemed able thereafter to manufacture the chemical. The phage hereditary material had evidently made good the deficiency in the human hereditary material! It will take many giant steps to proceed from this single, isolated experiment to the wholesale repair of hereditary defects by means of the bacteriophages. But we have at least had a glimpse of a whole new area of exciting medical possibilities.

Let us now turn from the bacteriophage-bacterium association to have a look at some less exotic microbial relationships. The protozoans commonly play host to both bacteria and algae. Some slipper-shaped paramecia appear green from the numerous green algae that they harbor within their single-celled body. These algae, like all green plants, manufacture their own food; in fact, they make much more than enough for their needs. And so, given adequate light, they can keep their protozoan host alive even if it gets virtually nothing else to eat. But if the algae—a species of *Chlorella*—are removed from the paramecium, the protozoan can survive only if it finds additional nutrients. So this is a mutually beneficial association. In return for shelter and transport toward light conditions favorable for photosynthesis, the algae feed their host.

It is a curious fact that the green paramecium eats free-living *Chlorella* algae, yet never harms those inside it. If a paramecium that harbors no algae is brought into contact with some free-living ones of the right species, the algae are taken into the protozoan, and there they multiply—but only until the host has its full complement of permanent guests. The multiplication then stops. Thereafter, if the paramecium meets any free-living *Chlorella* algae, it promptly swallows and digests them. Clearly, the protozoan is able to distinguish between its own algae and others, but nobody yet knows how.

Among the more striking types of association between protozoans and other small creatures, consider the complicated household of *Myxotrichia paradoxa,* a pear-shaped protozoan that lives in close association with four other kinds of living creatures: one insect and three other microorganisms. *Myxotrichia* itself is a guest in the intestine of a certain Australian termite, where it contributes to the insect's survival by helping it digest the pulverized wood on which it lives. A number of long flagella are attached to the narrower end of *Myxotrichia*'s cell, and the cell surface is covered by what early observers thought was a mass of shorter flagella. It took a detailed examination to reveal that these shorter hairlike "flagella" are actually long, twisty *spirochetes,* which belong to the bacterial group that includes the microbes responsible for the venereal disease syphilis.

Not only is the surface of the protozoan's cell covered with these creatures, but a large number of smaller bacteria cluster around the spirochetes, and yet a third kind of bacterium lives inside the cell. The spirochetes undoubtedly helps to propel the *Myxotrichia* along, even though it has flagella of its own. Otherwise, it is hard to determine the various benefits that each of these several types of microorganism gets from this web of associations. Quite probably, however, accessibility to food, as well as help in digestion, are among them.

Not many households within the invisible world are as complex as that of *Myxotrichia*. Even superficially simple relationships, though, are sometimes hard to fathom. For example, among the several types of association between different kinds of alga, there is one in which the host cells are green algae that have somehow lost their green pigmentation. Inside these cells, however, are threadlike filaments of a blue-green alga that appears to have taken over the food-manufacturing function of the host's lost green pigments. How does such a relationship develop? The most likely answer is that the host algae probably engulf the blue-green algae as food, but that the "meal" somehow manages to avoid being digested. Instead it apparently remains intact as a permanent source of nourishment.

Sometimes we can only guess at the respective roles of partners in some associations. Yet from time to time we do find new solutions to old mysteries. It never seemed clear what the bacteria contribute to most bacterium-protozoan associations, until scientists discovered that many such associations may be absolutely essential for the protozoan host. We have recently learned that the bacteria that live within at least one flagellate protozoan actually manufacture some of the amino acids that the host cannot make itself. And the amino acids are the basic building-blocks for all proteins.

We have seen that protozoans maintain intimate relationships with green algal cells that provide food for their hosts; with bacterial cells, which perhaps manufacture vital substances; and with spirochetes that are almost indistinguishable from the protozoan's own flagella. All of this lends support to the theory that the living cell of today, whether plant or animal, with all its complex internal structures may have arisen from much simpler ancestral cells that formed increasingly complex associations with other simple organisms.

It now seems probable, for example, that the *mitochondrion,* an ovoid structure found in virtually all cells and responsible for the production of energy, was once a free-living bacterium. Over millions of years, the theory goes, it established an association with the creatures that eventually evolved into today's plant and animal cells—an association that became so intimate that it was maintained and continued even when the cells reproduced.

Many people also believe that flagella and cilia were acquired in a similar way. Such microorganisms as *Myxotrichia* suggest how this might have happened. (In higher animals, some flagella and cilia have been drastically modified to perform utterly new functions. For instance, our own human cells that receive light and detect odors may well have been cilia or flagella in the distant past. And is it not probable that the chloroplasts that contain the pigment chlorophyll in all green plants were derived in some way from early forms of free-living algae?

It is difficult to find conclusive proof for any such theories, and it is indeed unlikely that we shall ever know the full evolutionary story. But the general hypothesis remains highly credible. It seems reasonable to believe that the structures of all the cells that make up the many varieties of plants and animals in the world today are the results of associations among microorganisms hundreds of millions of years ago.

Now let us look at the relationships involving microorganisms and insects. Such relationships are among the most noteworthy of all symbiotic associations. We have already seen how some protozoans live in the intestines of termites. Although termites eat wood, they cannot digest it and are entirely dependent on their microbial partners for this service. The digestive system of the insect is modified to provide a special chamber in which to house the protozoans.

Even more remarkable is the association of certain fungi with the insect world. Some of the relationships are quite simple, others extremely sophisticated. As an example of the simpler sort of relationship—one where the fungus merely lives in the insect's nest—let us take a look at what happens in plant galls. A plant gall is a swelling that sometimes appears on a stem, leaf, or bud, often as a result of insects' having deposited their eggs in the plant. In some way that is not understood, this causes the formation of abnormal plant tissue, which nourishes the insect larvae when they emerge from the eggs. Galls caused by some mosquitolike insects contain fungi, which form a thick layer inside the gall. It is possible that the female insect

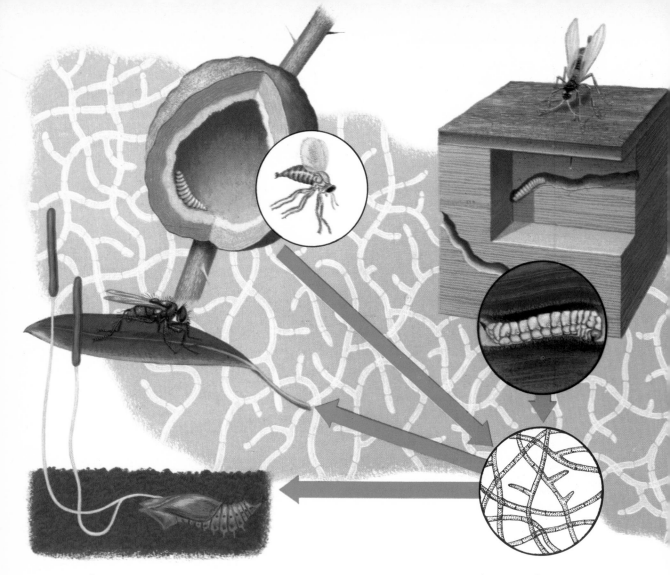

Insects and Fungi

1 The fungus *Cordiceps* parasitizing a buried moth pupa.
2 A fly parasitized by the fungus *Entomophora*. A network of hyphae eventually kills the host.
3 An adult gall midge. When the female deposits eggs in a plant a hollow gall develops, and fungal spores laid with the eggs form a layer of fungus on the inside.
4 Gall section showing a midge larva feeding on the fungus layer.
5 A female wood wasp laying eggs by means of a needle-like ovipositor. Fungal spores deposited with the eggs produce hyphae that ramify through the wood.
6 A wood wasp larva. The developing larva burrows into the wood feeding on the fungus as it goes.

7 An ambrosia beetle, feeding on the fungal mycelium, or "ambrosia," that lines its tunnel walls. The fungus develops a yeast-like shape.
8 The pattern of tunnels made by ambrosia beetles in wood. The "cul-de-sacs" house beetle larvae.
9 Ants carrying cut leaves to their nest as compost for fungus gardens.
10 The leaf-cutting ant's nest, showing the fungus gardens.
11 An ant tending a fungus garden. Constant pruning causes the fungus to develop swollen tips.
12 A fungus-growing termite mound with fungus gardens and fruiting bodies.
13 A termite in a fungus garden. Constant "gardening" causes the fungus to produce white balls, which form the termite's food.

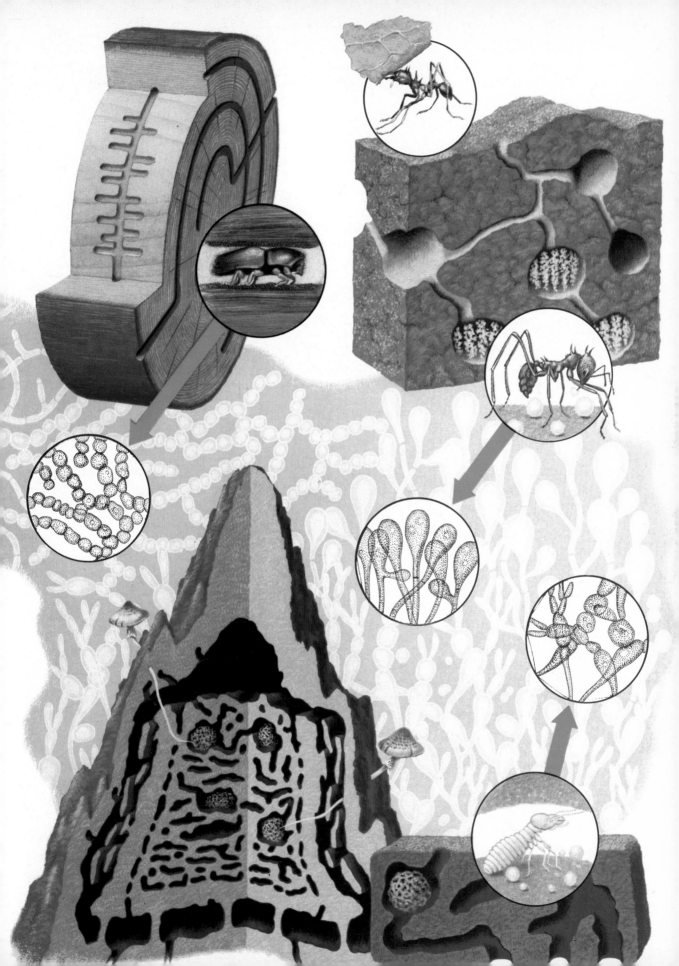

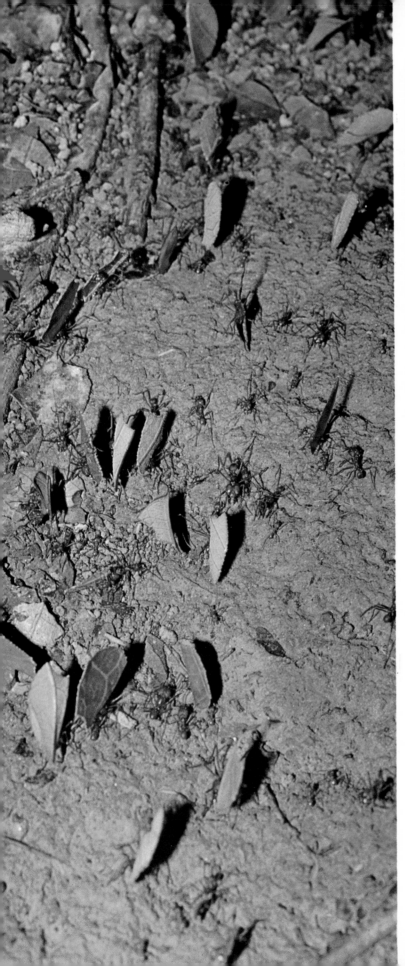

The leaf-cutter ants (left and below) use chewed-up leaf fragments as compost on which to grow their fungus food, whereas the fungus in the opened larval nest or gall (above) is thought to make the gall's woody tissues more digestible for its insect larvae.

Above: termites foraging for food. Termites digest grass, leaves, dead wood, and other materials containing cellulose with the aid of protozoa (below) in their gut. Some species of termite supplement their diet by cultivating fungi in special chambers within their nests (right). These are called fungal gardens.

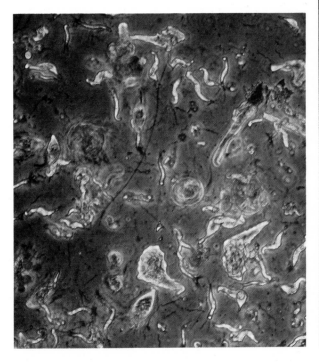

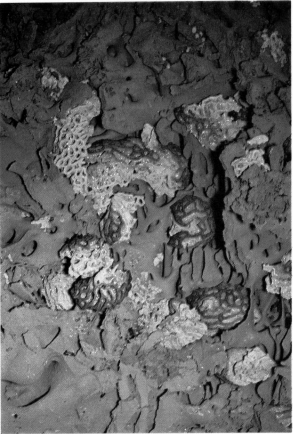

deposits the spores of the fungus at the same time as she lays her eggs. We do not really know what role the fungus plays in the gall. It is unlikely that the larvae feed on it directly; more probably the fungus helps to break down the gall tissue, making it more easily digestible for the young insects.

The gall example illustrates a comparatively loose ecological relationship. A much closer one is enjoyed by another type of fungus and the scale insects that inhabit it. The tiny fungal microorganisms colonize and create a thick lichenlike mat on the leaves and branches of trees, and this habitable mass of fine threads provides shelter from the weather and protection from predatory birds for the scale insects. The insect gets its nourishment from the tree by tapping the sap-carrying veins with a long sucking tube, and the fungus extracts food from insect blood by sending out fine threads to penetrate the bodies of a few of the little animals. The insects invariably suck out more sap than they need, and there is plenty of nourishment for both animal and fungus.

The fungus gets another and equally important

benefit from the association. Young scale insects become contaminated with fungal spores while crawling upon the fungal mat; and when they migrate to fresh parts of the tree, they carry the spores with them. These, of course, are the starting points for new fungal growth. Thus the fungus provides shelter for the insects, and the insects provide both food and a means of dispersal for the fungus.

Another animal that rewards its fungal associates for services rendered by helping to disperse it is the wood wasp. The female wasp lays her eggs in damp wood by means of a long, slender ovipositor, at the base of which are tiny pouches containing fungal cells. The cells cling to the deposited eggs and give rise to a mass of fungal threads, which, as they grow into the wood, leave channels filled with fungi. When the larvae hatch, they move along these channels, eating the fungus as they go. The fungus, which feeds on the wood, does not need the wasp for food, only for dispersal; the wasp, however, needs the fungus to help it break down and digest the wood, which it cannot assimilate alone.

Female wood-wasp larvae have small pouches

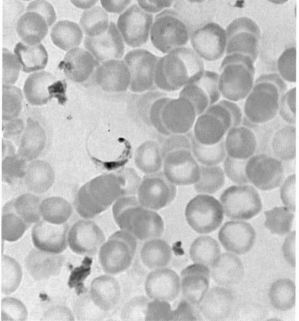

Minute single-celled organisms, such as Trypanosoma *(above), are responsible for a large number of diseases in man. Many of these organisms live part of their life-cycle in insects, part in man and other animals. One of the insect carriers of* Trypanosoma, *which causes anemia and fever in man, is the blood-sucking South American Rhodnius bug (right). The human louse, (left), is the carrier of the organism responsible for typhus*—Rickettsia.

tucked away in a fold between the first and second segments, and small pieces of fungus are trapped by a waxy substance inside the pouches. When the larva becomes a pupa, it loses its pouches, but the adult female picks up flakes of fungus-impregnated wax as she emerges from the pupal skin, and it is these that become lodged in the pouches at the base of the ovipositor. Thus the transfer of the fungus from one generation of wasps to the next is assured.

Some insects actually maintain fungus "gardens"—an outstandingly sophisticated type of microorganism interaction. The ambrosia beetle does this, for example; in small pockets in its hard outer skeleton this wood-boring insect carries some fungal spores. Ambrosia beetles attack hardwood and fallen timber that is moist and filled with sap (Europeans call them "beer beetles" because they sometimes bore into beer and wine kegs). As the beetle bores its way through the wood, fungal spores spill out of the pockets and germinate, so that the tunnel becomes lined with a mass of velvety fungus. The fungal cells feed on the wood of the tunnel, and the fungal mycelium forms the "ambrosia" on

which the insect feeds, and for which it is named.

You can distinguish the borings of an ambrosia beetle from those of other beetle species by the black or brown discoloration of the fungal spores that surround the neat, circular tunnel opening. The tunnel contains the beetle's larvae too—either in separate niches or in communal chambers. When the adult insect emerges from the pupa, it feeds on the mass of ambrosial fungus that lines its part of the tunnel; as it eats, it rocks backward and forward, thus making sure that its pockets become filled with fungal spores.

Perhaps the most interesting examples of fungal gardens are the ones tenderly cultivated by certain tropical termites—destructive creatures that do an estimated annual amount of damage to West African buildings equal to 10 per cent of the value of the buildings attacked. These termites invade all manner of materials, from wood to rubber, from growing crops to underground cables. Their nests may be huge mounds of hardened earth or may be completely subterranean. Each such nest contains one or more fungus gardens, which look rather like gray or brownish sponge, often convoluted like

the flesh of a walnut, and which may be very large indeed. A typical garden can measure as much as two feet in diameter and weigh up to 60 pounds. The garden is always enclosed tightly in a cavity lined with a mixture of saliva and dirt; and some species of termites even ventilate their gardens by means of an elaborate system of vertical conduits that extend to the surface of the nest.

The insects tend the gardens diligently. They even fertilize them with manure composed of the partly digested feces of worker termites. And they "weed" the gardens, too, by removing bacteria and alien fungal spores. The termites continually eat away at their gardens, and it seems probable that the fungi help them to digest other materials that they have swallowed; possibly the microorganisms also provide vita-

mins for the insects. The fungus flourishes as a result of its rich diet of termite excrement.

Certain ants, too, are keen fungus gardeners. Among these are the leaf-cutting ants, found only in some of the hotter parts of the Western Hemisphere. One species, which lives in eastern Texas and southern Louisiana, builds nests that may be 50 feet across and 20 feet deep. New nests are initiated by a young queen ant, which carries a small pellet of fungus in a pouch just below her mouthparts. She uses this for starting up a small fungus garden on her excreta, and feeds the fungal matter to the first larvae of the worker ants. When they are mature, the workers forage for such foods as caterpillar excrement, fallen flower parts, and other soft plant debris, and they also cut leaves from trees. Unlike the termite gardeners, these ants do not begin the

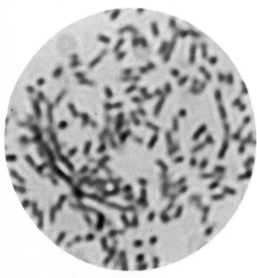

When the Flemish painter Pieter Bruegel (1520–69) painted The Triumph of Death *(right) he was portraying a religious message from the Apocalypse. But he was relying for impact on the fears of the terrible Black Death of 200 years earlier, when some 25 million people died. The cause of the Black Death, and its recurrent outbreaks was the plague bacteria (left). These were carried to man by fleas (below left) as they left their first victims—man's unloved companions, the rats.*

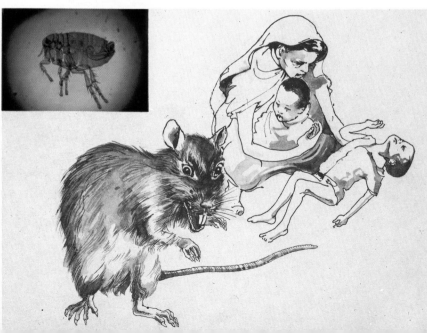

process of digestion by swallowing the results of their foragings. Instead, they add their findings, chopped into small pieces, directly to the garden—which, in consequence, looks like a gray mass of finely cut bits of plant material held together by threads of fungus.

The adult ants feed on the swollen tips of the threads. These are found in clusters, which, to the naked eye, are barely visible as minute white dots in the garden. (The clusters are known as "kohlrabi" bodies, but they are not to be confused with the vegetable of the same name.) As long as the ants are actively tending their garden, the fungi do not produce fruiting bodies, although they start sending up mushrooms as soon as a nest is abandoned. How the ants impose this sort of birth control on their fungi is not known. Perhaps constant "pruning" inhibits

the development of spore-dispersal structures.

In some insects, bacteria or yeasts actually grow within the body cells, which become specialized for just this function. Such cells are called *mycetocytes,* and although they may be found scattered throughout the body of an insect, they are more often combined together to form special organs. Frequently these organs are merely blind sacs opening into the gut, and the reason for their existence is nutritional. When such insects are deprived of their internal guests, they invariably suffer from lack of vitamins, and sometimes from protein deficiencies as well. Thus, microorganisms are welcome guests in numerous ecological associations because they either help their hosts to digest substances that the hosts cannot tackle, or make substances required for the hosts' well-being.

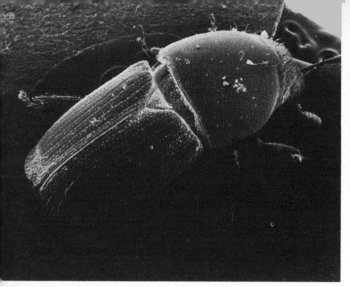

Right: an elm killed by Dutch elm disease. Above: a cross section of an elm-tree stem showing the black spots of fungus that killed the tree and, left, a bark-boring insect that lives on the wood of the elm tree and transmits the fungus.

The microorganisms that live in or on insects often have a significance that goes beyond the biologically interesting nature of the association, because some insects carry disease, and the diseases they carry are usually caused by their invisible partners. The mosquito and many other insects that carry disease-producing microbes to their victims in the course of pursuing their own life activities are said to be *vectors*. There is a wide range of possible associations between a microorganism and its vector. In some cases the relationship is simple and unspecific; in others, the microorganism, the insect vector, and the ultimate victim are essential parts of a complex cycle of infection and disease.

The common housefly is a familiar example of the simple type of relationship. Houseflies are not in themselves harmful, nor do they have any sort of specific association with a harmful microbe. But in their search for food they crawl over decaying matter—often excreta of other animals—which is sure to be a paradise for microorganisms. The microbes stick to the fly's rough skin and to the hairy pads on its feet. Then, when it alights on human food, the microorganisms are transferred to the food, and thence to some hapless victim. Enteric bacteria, which cause diseases of the intestine, are particularly likely to be spread in this way.

We find a much more complex pattern of association when we look for the genesis of rickettsial diseases—those brought on by a bacterium-like organism (which may really be a protozoan) belonging to the genus *Rickettsia*. The rickettsiae cause diseases ranging from Rocky Mountain spotted fever to a mysterious ailment called "Q-fever," but they are best known for causing typhus, a very infectious and often lethal disease.

(Typhus, by the way, is not the same as typhoid, for which a rickettsia is not responsible.) The vector for typhus is a louse—either the human head louse or the body louse. The rickettsias live in the gut of the louse and are liberated in its feces. When the louse bites a person, it may simultaneously defecate onto the punctured skin, thus almost inevitably providing the microorganisms with an entry into the body. If an uninfected louse bites the infected victim, it picks up some rickettsias with its meal of blood, and so can transmit the typhus to another.

Ticks and mites also help to spread rickettsial diseases. Rocky Mountain spotted fever is carried by ticks, which transfer the microorganisms to man. An ailment with the unpleasant name of "rickettsial pox" is carried by mites, which can transfer it to mice or people.

A particularly nasty series of associations is responsible for the dreaded Black Death (or bubonic plague), which destroyed one quarter of Europe's population in the 1300s, and which decimated the population of London in 1665. *Pasteurella pestis,* the bacterium that causes the disease, lives in the bloodstream of rats and is passed along to rat fleas when they suck their host's blood. These fleas attack human beings as well as rats, and so they transmit the plague further in rat-infested places. Bubonic plague remains a menace in some parts of Asia and has been a very serious problem in war-ridden Vietnam during recent years.

Among the other harmful microorganisms passed along by insect vectors, one of the most virulent is the thin, crescent-shaped protozoan that causes sleeping sickness. It is transmitted to man by the tsetse fly, a bloodsucking fly found only in Africa. The protozoan lives initially in

the gut of the fly, but it soon invades the salivary glands and the mouthparts, from where it is injected into a human host when the fly takes a meal. In man it lives and grows chiefly in the bloodstream, but in the latter stages of the disease it invades the brain and spinal cord.

The viruses are less likely than protozoans to be passed along to man by infected insects, although there *are* a few viral diseases that human beings can catch in this way. Fleas, however, do carry one of the best-known viral diseases of animals, myxomatosis, which killed most of western Europe's rabbits a few years ago. There are very specific associations here between both the virus and its host flea and the flea and its host. When an infected rabbit dies in its crowded burrow, the fleas quickly leave the dead body, in search of other hosts. When they find a new host they expel their viral parasites into the victim's body through their mouthparts. Curiously enough, though, there is a less virulent form of this virus that is carried by a mosquito. A rabbit bitten by an infected mosquito is in the same category as a person inoculated with cowpox: it becomes immune to the more dangerous disease because the less harmful disease calls forth the right antibodies.

Insect vectors carry destructive microorganisms to plants as well as to animals. Dutch elm disease, a modern scourge that threatens the entire elm population of both the Old and the New World, is caused by a fungus that effectively starves trees by blocking up the channels through which the plant's food and water must flow. This fungus is transmitted by bark-boring beetles, which live on elm wood and carry fungal spores about with them in much the same fashion as do the ambrosia beetles. By cultivating fungal gardens in their brood tunnels, the bark beetles bring death to the elms, which have no built-in defenses against their killers.

Insects are particularly important as vectors in the spread of viral diseases of plants. Such diseases often result from a very simple form of association in which an insect with biting mouthparts feeds on an infected plant and carries away the virus smeared on its proboscis. Thus it becomes a kind of "unwitting" vector. Such casual associations are typical of a number of different species of aphid. The aphids are probably responsible for transmitting the great majority of plant viral diseases; one species alone is the vector of 50 different viruses. Of several major types of aphid-virus association, the simplest consists of the mechanical transmission of the virus from one plant to another on the mouthparts of the aphid. No biological association is involved, and usually only the first

plant attacked is actually infected by the virus.

More complex is the "circulative" type of behavior. Here the aphid again merely picks up a virus from a diseased plant, but another plant cannot be infected until the virus has passed through the insect's gut, into its bloodstream, and back to the salivary glands.

A third and much more complex pattern involves the so-called "propagative" virus, which has a definite biological association with its host and lives and multiplies within the aphid's body. An example is the virus that causes leaf roll in a number of plants. An aphid infected with leaf-roll virus is able to pass it on to a susceptible plant even after seven days of feeding on a plant such as cabbage, which is immune to leaf roll, whereas other types of virus would have been destroyed in the insect's body in that time.

In some plant diseases there is cooperation among three creatures—an aphid and two viruses. One such three-way association is responsible for "rosette disease," which results in the discoloration and deformation of tobacco plants. In this ailment, one microorganism (the mottle virus) depends on the presence of another (the vein-distorting virus) to enable it to be picked up and transmitted by an aphid. Precisely what this sort of cooperation involves is as yet a mystery to microbiologists.

Plant virus diseases are spread not only by bark beetles and aphids, but also by many other creatures, such as mites, eelworms, and even fungi, but we do not always know *how*.

The vectors themselves do not always get off scot-free, of course, for they are frequently struck down by their invisible associates. For example, there is one bacterium that produces a poisonous protein that crystallizes around its spores, with lethal results for the larvae of butterflies and moths. Larvae that swallow these bacteria are quickly paralyzed, for the poison drastically affects the ability of the larva's intestine to absorb food and water. The bacterial spores then germinate and feed on their unfortunate hosts, which soon die.

So effective is this poison against over 100 species of moth that it is now being used as an insecticide. The bacteria are grown in large fermentation vessels, and a concentrate consisting of spores and poisonous crystals is prepared and used as a dust or spray on larvae-infested plants. This concentrate seems to be harmless to human beings and livestock. Some authorities believe that there may be great possibilities in the use of other such poison-making bacteria as a new approach to insect control.

About 200 viral diseases of insects have been identified, all of them affecting chiefly the

There are many examples of the beneficial associations between marine creatures and microorganisms. One of the most startling is the association between certain fish and luminous bacteria. The angler fish (above), for instance, uses luminous bacteria to lure its prey into its jaws. Right: the blue coloration in this close-up of the mantle of the giant Tridacnid clam, found on Australia's Great Barrier Reef, is made up of colonies of algae living in the clam's tissues. Light is concentrated on to the chlorophyll-rich algae by the many tiny lenses in the mantle. As the algae multiply they are absorbed as a food supplement by the clam. Far right: photomicrograph showing green alga that has lost its chlorophyll-producing cells. Filaments of the blue-green algae can be seen apparently substituting for the host's missing cells.

larval stages. The larvae may become infected by feeding on plants contaminated by other larvae that have been killed by viruses. There are also many protozoan-engendered and fungal ailments of adult insects.

Before we move on to examine in the next chapter the many kinds of association that exist among microorganisms and mammals, let us glance briefly at just two or three of the interesting living relationships with some creatures not yet mentioned—specifically, with one bird and with a few marine organisms.

Certain birds that are native to Africa and India are called honey-guide birds because they guide ratels, or honey badgers, and other animals—or people, for that matter—to bees' nests. The bird's aim is to have the nest ransacked—not, strangely enough, so that it can have a share of the honey, but so that, after the larger animal has helped itself to the honey, the

bird can eat the broken remains of the wax honeycomb. Even more strangely, the honey-guide bird cannot digest the wax it craves; it must rely on the presence in its alimentary tract of two kinds of microorganism, one a bacterium, the other a yeast. Working in partnership, these microorganisms are able to break down the wax and render the products available to the bird. The delicacy of the association can be judged from the fact that the bacterium could not do its part of the breaking-down job effectively without the help of a certain chemical substance produced by the bird and present in its alimentary canal.

And so from land to water, with a quick look at some examples of aquatic associations. More than 100 different kinds of invertebrate, including jellyfish, corals, sea anemones, flatworms, sponges, and clams, are known to have symbiotic relationships with algae. One very

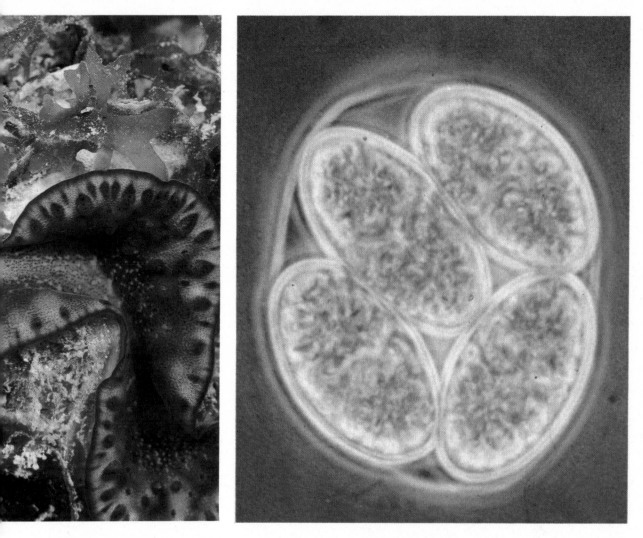

specialized association, found on the Great Barrier Reef of Australia, is the relationship between a type of large clam, the tridacnid clam, and certain yellow or greenish brown algae.

Tridacnid clams have much thicker layers of tissues lining the two halves of their shells than clams usually do, and the algae are concentrated in these tissues. Groups of cells that contain the algae are arranged into conical protuberances that are associated with transparent cells. These transparent cells act as lenses, whose purpose is to focus daylight onto the chlorophyll-rich algal cells so as to enable them to make food. As the algal population in the clam grows, the clam roots out and digests the excess cells.

Some aquatic animals harbor microorganisms for quite unusual purposes. Cuttlefish, some squids, and a number of fishes have special organs that house luminous bacteria. For deep-sea fishes, the light emitted by the bacteria is no doubt an assistance to navigation in the dark water; other species use the light as a signal for mutual recognition. One species of squid is even able to control the intensity of the light, because the organ that houses the bacteria is embedded in the squid's ink sac and partly enclosed by reflective tissues. By changing the position of the dense ink sac in relation to the reflective tissue, the squid can darken or brighten the light. In the luminous fishes, bacteria-housing organs are found under the eyes, on the lower jaw, on the belly, or around the rectum.

In this chapter we have seen microorganisms at work as biochemical warriors (bacteriophages attacking bacteria), digestive aids (gut bacteria of insects), and living headlights (luminescent bacteria in fish) among a host of such other roles as carriers of disease. All these examples are evidence of the amazing variety of parts the microbes play to ensure their survival.

Microbes and Mammals

Healthy children at play are a commonplace sight in our present-day Western world. Under a century ago, the lethargy of the gravely ill child was just as familiar. The transformation largely reflects mounting medical knowledge of and mastery over the microbes that flourish unseen both on and in our bodies.

Human beings are self-centered creatures. Our own lives, the things we do to fill them, and our health are what matters most to us, and any threat to our well-being provides the best possible impetus for us to search for the source of the trouble and try to root it out. Disease is just such a threat. Ever since it became clear in the middle of the 19th century that microorganisms are responsible for most diseases, scientists have mounted an extensive campaign to discover why various organisms cause various diseases and how to combat them. Of all associations

between microorganisms and animals, therefore, we know most about those that involve man.

It is natural to think of microorganisms chiefly as an agent of disease—as "germs"—but this is only a small part of the complete story. Healthy animals (including man) play host to a wide range of microorganisms, some potentially dangerous and some not, without developing symptoms of disease. In fact, it is often very hard to determine whether the relationship between a microorganism and a higher animal is harmful, beneficial, or neutral. It all depends on the circumstances. In your nose and throat, for example, there are almost certainly microorganisms that can cause diphtheria or pneumonia. That they do not do so except in unusual circumstances (when the body's defenses are weakened, for instance) is a tribute to the remarkable system of checks and balances that enables man and microbe to live in close harmony.

In healthy individuals, microorganisms are chiefly found living on exposed parts of the body. The skin is quite obviously one such region—although, as we shall see, the actual surface of

the skin is less hospitable to microlife than are its pores and crevices. Another, less obvious habitat is the linings of the digestive system, which, like those of the tubes leading from the nose to the lungs, are also exposed, because the digestive system is merely a tube running through the body from mouth to anus. Thus, the contents of the tube remain effectively outside the body unless they are absorbed through the wall of the tube.

For many varieties of microorganism, the digestive system is the most suitable of all places to live, providing, as it does, a steady supply of food for its microscopic residents. Furthermore, the temperature and amount of moisture present in the digestive system of a healthy animal are ideal for microbial growth. Small

wonder, then, that no other single habitat contains such a concentration of microorganisms of so many different kinds. "Single habitat" is perhaps a misleading term, because a mammal's digestive system encompasses a huge variety of different habitats, each harboring its own specialized kind of microbe. In studying the ecology of plants and higher animals, the separation between biologically different habitats is likely to be measured in yards or miles. But the distance between two different habitats that support entirely distinct kinds of microorganism may be only a few thousandths of an inch.

Despite the apparently easy living that the digestive system seems to offer, it is not suited to all species of microbe, for mammals have a

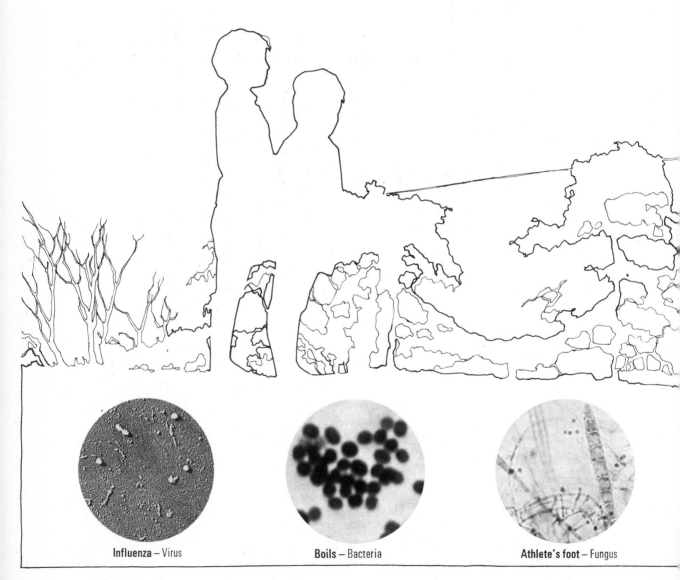

Influenza – Virus

Boils – Bacteria

Athlete's foot – Fungus

range of defense mechanisms that work together to prevent the invasion of harmful microorganisms as well as to control the number of those allowed to take up residence. Among the defenses is the presence of special cells that can swallow up and destroy microbial trespassers, and the body also produces chemical substances especially designed to kill off intrusive forms of life. We shall examine these defenses more carefully when we take a closer look at disease-causing microbes.

Only such organisms as can overcome or come to terms with the body's defensive mechanisms are able successfully to colonize man and other mammals. But do not assume that all the microorganisms that live in or on higher animals must continually fight to sustain their position.

The microbiologist's microscope puts human life in a new and unfamiliar perspective. Through this instrument we learn that flesh and blood play host to a miniature world of wildlife—some of it hostile, some neutral, some even beneficial to us. Below are photomicrographs of a few kinds of "enemy" germ that can thrive on or in the body and create disease if they multiply sufficiently. Most bring mere discomfort, but at least one kind is often fatal. Fungi producing ringworm and athlete's foot attack the skin: common ringworm shows as red, round, often itchy patches on non-hairy skin; athlete's foot is an inflammation and cracking between toes. Staphylococcus bacteria entering a sweat gland may form a boil—a painful raised round abscess. Tetanus bacteria penetrating the skin through an open cut give off poisons and trigger severe muscle spasms that can kill if left untreated. Yet other disease organisms are breathed in with droplets in the air. These germs spread down through the body's respiration system, then get into the bloodstream, causing general malaise. Two such kinds of germ are the viruses responsible for influenza and measles. Headache, shivering, and aching are common influenza symptoms. Measles produces a spotty rash, sniffles, a high temperature, and exhaustion.

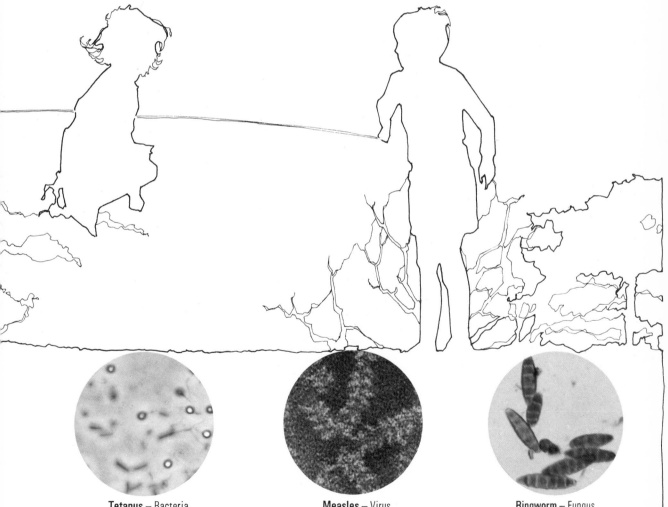

Tetanus – Bacteria   **Measles** – Virus   **Ringworm** – Fungus

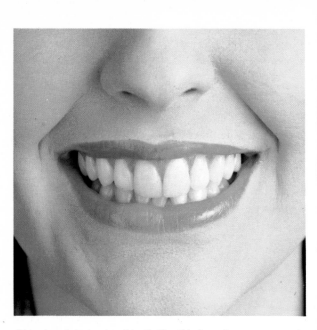

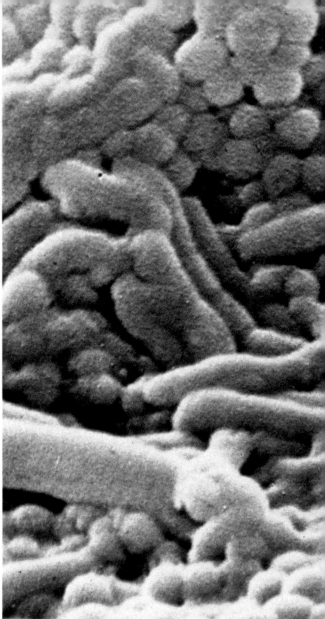

Gleaming, intact sets of teeth like this have become scarce in "developed" nations where sugary foods encourage bacteria whose acid wastes cause tooth decay. Right (greatly enlarged): the branching, knobbly surface of plaque—a bacterial mass found on teeth. Plaque can be kept at bay by frequently brushing the teeth and probing between them with a toothpick.

Some are so beneficial to the host that it has developed ways of encouraging a peaceful and mutually profitable coexistence.

Each region of the digestive system has a distinctive array of microbial inhabitants. First, there is the mouth. Here the immediate threat to the survival of microbial colonizers is the saliva, which, apart from being poor in the kinds of substances on which microorganisms can feed, contains chemicals that can kill them. The teeth appear to be the most important factor in determining what kinds of bacterium are found in an animal's mouth, for teeth provide firm surfaces on which food particles collect and microbes can grow; and the teeth are not in themselves able to attack microorganisms either chemically or physically. Bacteria on the teeth form a tenacious coating—the so-called dental *plaque*—which resists even hard brushing. Plaque is constructed of fine, threadlike bacteria woven together into a dense mat, which in unbrushed places can be as much as 60 times the thickness of an individual bacterial cell.

Bacteria are not the only microorganisms that live in the mouth; it is commonly inhabited by protozoans and fungi as well, including an amoebalike protozoan that lives in and around the gums and feeds on the bacteria it finds there. But bacteria are the chief factor in tooth decay. They gather in the cracks and crevices spared by the toothbrush and the self-cleaning action of tongue and lips, and there they feed on particles of food. They break these down chemically, deriving energy for themselves, but the products of the process, a variety of acidic substances, attack the enamel tooth surfaces.

Once the hard enamel has been breached, further chemical substances produced by bacteria insidiously destroy the inner matter of the teeth. Enamel that contains the chemical fluoride as part of its structure is particularly resistant to the acids formed by bacteria, and this is why many dentists maintain that fluoride should be added to drinking water. The teeth of

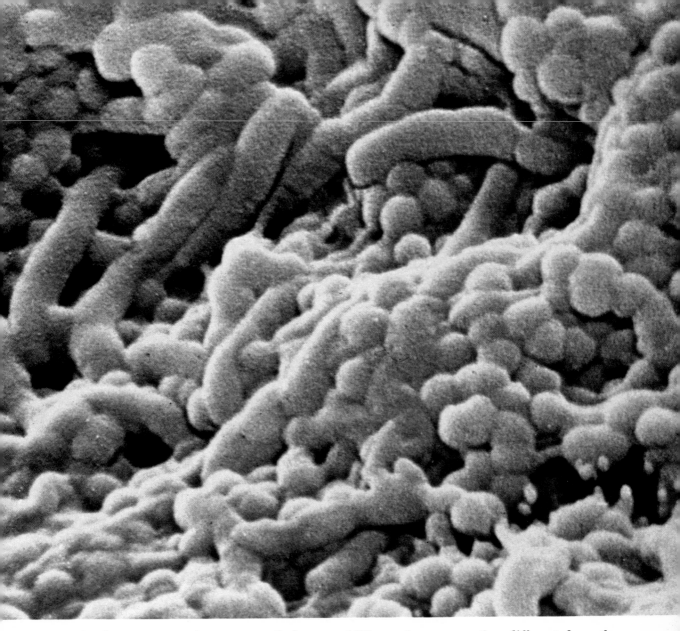

some animals are more resistant to tooth decay than are those of human beings; dogs, for example, have few dental problems because their teeth are shaped in such a way that food particles cannot lodge in the crevices.

It is. an interesting fact that the threadlike bacteria of dental plaque are not found in the mouths of human babies during the toothless early months, but take up residence as soon as the teeth begin to appear. Babies in the womb are completely free from bacteria; yet within hours of birth they have the beginnings of a characteristic bacterial population. The baby gets its microorganisms first from its mother during the passage down the birth canal and later by contact with the world outside. The bacterial populations in the gut of the newborn

child are, however, rather different from those that develop later on—a reflection of changing diets. The digestive system of breast-fed infants contains only one type of bacterium, whereas bottle-fed infants are hosts to quite a large range of microorganisms. As the child grows older and its diet changes, the kinds of micro-organisms that its digestive system will support also increase and change. But we still know very little about how even the commonest of micro-organisms have adapted to survive in the digestive tract.

Partly digested and macerated food passes from the mouth to the stomach, which contains acids strong enough to blister a human hand—and certainly strong enough to discourage the growth of most bacteria. The stomach is therefore

The quantities and kinds of microorganism that live in a man's digestive system change from human infancy to adulthood. Newborn babies are largely germ free, and acid in the digestive systems of breast-fed babies helps to kill off microbes. But a bottle-fed baby (left) soon acquires a variety of intestinal bacteria—many sucked in from the baby's food container. By adulthood, the human gut has a flora of usually harmless and beneficial bacteria. Densest concentrations flourish in the large intestine (part of which appears in section in the photomicrograph above). There, the bacteria yield gases and other wastes by breaking down food substances that had escaped absorption by the small intestine. More importantly for man, certain bacteria in the large intestine produce nutritionally valuable quantities of riboflavin and other vitamins.

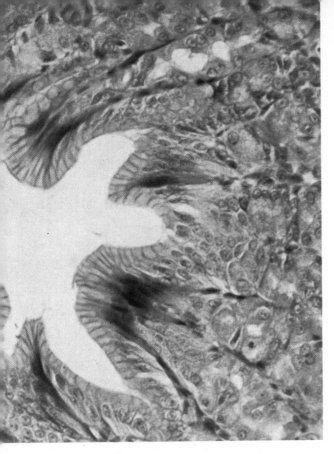

an effective barrier, normally preventing dangerous bacteria from gaining access to the intestines. Nevertheless, large numbers of acid-resistant bacteria and a few fungi are able to make their home in the stomach. And when its normal condition is disturbed—in cases of stomach cancer, for example—it may become less acidic, and thus may allow the free growth of yeasts and bacteria.

Where the small intestine leaves the stomach, it too is very acidic, and so the kinds of microorganisms found there are much like those of the stomach. But the acidity declines and the number of microorganisms increases toward the lower end of the small intestine, just before the beginning of the large intestine. If the stomach is a barrier to microorganisms, the large intestine is a paradise; they are there in enormous numbers. Some live within the mass of digesting food that passes through the intestine, and these microorganisms are eventually expelled along with the fecal material, every ounce of which can contain as many as 300,000 million of the tiny creatures.

But if the bacterial population of the large intestine is to remain fairly constant, the millions of bacteria voided with the feces must be replaced. Here the microbes' salvation is their

speed of reproduction. Food takes about 24 hours to pass through the digestive system; in that time the bacteria in the digesting mass may double or quadruple their numbers.

One type of bacterium, which inhabits the inner wall of the intestine, has a distinctive, elongated, threadlike shape, which you would easily recognize if you examined a portion of the large intestine under a microscope. These bacteria do not need oxygen to live (and this is just as well, for there is none in the large intestine). There are also likely to be a few species of protozoan in the digestive tract of a healthy human being. One such protozoan, *Entamoeba histolytica,* appears to thrive in the intestines of a high percentage of people who live in places where sanitation is poor. In most cases it seems to do no harm. In a few individuals, however, it damages the intestines and leads to an unpleasant disease called *amoebic dysentery.*

For a dramatic example of how such an association can benefit the host instead of doing damage, consider the digestive systems of such grass-eaters (herbivores) as the horse, pig, porcupine, rabbit, and guinea pig. These mammals, which depend on green plants for their nourishment, cannot digest cellulose, the tough material that forms plant cell walls. They have overcome this problem by developing a special region of the intestine for the sole purpose of housing colonies of microorganisms that can break the cellulose into simpler substances that the host animal can use. This special organ—the *caecum*—is a saclike, blind extension of the large intestine, inhabited by bacteria and protozoans.

Foodstuffs pass down the intestine and into the caecum. Here the cellulose is broken down by the microorganisms, and the energy contained in the food thus becomes available to both the microbe and its host. One drawback for the host is that because the caecum is a blind alley situated toward the hind end of the gut, much of the foodstuff does not have a chance to be fully broken down by the bacteria; and so a good deal of valuable nourishment is excreted

Some herbivores, such as rabbits and guinea pigs, partly overcome this problem by eating their excrement—an interesting practice known as *coprophagy.* The fecal matter that has passed through their bodies only once is soft and moist. Wild rabbits void this during the day when they are below ground, then immediately eat it again and digest what nourishment remains.

During the second passage, most of the water is extracted. This is why the rabbit pellets that you see around grassy places where rabbits are common are hard and dry.

The most efficient and highly developed system for using microorganisms to supplement mammalian digestive processes—digestion by proxy, as it were—is to be found among *ruminants*: cattle, sheep, goats, and their relatives. These animals get their collective name from the *rumen,* which is a special organ that houses gut microorganisms. Because of the economic importance of cattle and sheep, rumen microbiology and biochemistry have been extensively studied, and we know a lot about the system.

To begin with, the rumen is an extension of the esophagus (the tube that connects the mouth to the stomach), which makes it a sort of first stomach—the first part of the digestive system to receive food after it has been swallowed. The rumen is very large, with a capacity of over 20 gallons in cows and over one gallon in sheep, and it has a complex structure. When food enters the rumen, it becomes mingled with the resident microbial population, consisting chiefly of bacteria and protozoans. The role of the protozoans—which seem to be mostly ciliates—is not fully understood. They apparently play some small part in digesting the food, but their chief function may well be related to the control of the bacteria in the rumen by gobbling up the surplus.

The rumen bacteria are fermenters; indeed, the rumen itself is a large fermentation chamber in which the bacteria convert cellulose into sugars, and then convert the sugars into a number of organic acids. These acids are absorbed through the wall of the rumen into the bloodstream and thus become the chief source of energy for the ruminant. This is a very different form of digestion from that of most mammals, whose digestive systems produce chemicals that convert carbohydrates in their diet into sugars for direct absorption into the bloodstream through the wall of the gut.

As the microorganisms break down the sugars into their component acids, they produce, as a by-product, large quantities of carbon dioxide and methane gas—as much as 45 gallons a day, in fact. And all this gas must be expelled by belching (a process sometimes delicately termed *eructation*). Belching in ruminant animals is more than a mere occasional hiccup. The walls of the rumen undergo two quite separate kinds

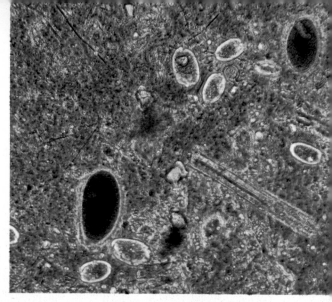

Countless tiny microbes like the protozoans in the photomicrograph above helped this cow to build up its large body. Inside its digestive system, bacteria (their numbers controlled partly by protozoan predators and partly by digestive juices) transform grass to simple energy-producing substances. The bacteria then serve as a body-building protein.

of contraction. One of these serves to mix the contents of the rumen together; the other, a forward-moving contraction, causes the animal to belch. If the gases are not released from the digestive system (as sometimes happens), the animal becomes bloated and dies.

Efficient as the rumen bacteria are at converting cellulose into nourishing acids, they cannot entirely finish the job. Coarse food materials that the microbes have not broken down are returned to the animal's mouth to be chewed again—which is the process of rumination ("chewing the cud"). What happens is that food particles not fully fermented by the bacteria scratch against the sensitive walls of the rumen, causing its upper end to open and permit a ball of undigested food to be sucked into the esophagus and up to the mouth. The food is once again chewed, swallowed, and worked upon by the microorganisms. After a short period, the entire process is repeated.

To keep healthy, mammals need not only carbohydrates for energy but also proteins, fats, and vitamins for growth. So far in our story the ruminant has been supplied only with the necessary carbohydrates by its bacterial partners. They, of course, have been using some of the energy they gain from the fermentation of cellulose to grow and reproduce, and their growth inevitably involves the manufacture of

proteins, fats, and vitamins. After fermentation in the rumen, the food is passed on to what in other mammals would be described as the true stomach. And just as in other mammals, the stomach acids kill most of the residual bacteria in the food mass. The mixture of partly digested food and dead bacteria passes into the intestines, and the animal absorbs what it needs from the mass. So, in fact, it is the *bacteria* that the ruminant digests which provide it with the essential vitamins, fats, and proteins.

In a sense, therefore, many herbivorous animals maintain their own herds of micro-organisms, allowing them to graze and fatten in an especially constructed pasture, and killing them off and digesting them after they have done their work. No similar system for breaking down cellulose exists in man, and we do not digest that part of whatever green vegetables we eat. Cellulose in our diet is chiefly of value as "roughage"—bulky material that helps to keep the muscles of the digestive system in trim.

Microorganisms are found in large numbers in the human nose and throat, in the passages leading to the urinary and reproductive organs, and on the skin. The outer surface of the skin is not a very comfortable place for the microbes, because it is exposed to the atmosphere, which dries it up, and most microbes thrive best in fairly moist habitats. Only the scalp, the areas of the face that bear hair, the passages leading from the ears to the ear drums, the underarm, urinary, and anal regions, and the spaces between the toes are sufficiently moist to support appreciable microbial populations. On this simple fact rests the multimillion-dollar deodorant industry.

There would be no need for deodorants if it were not for the association of skin micro-organisms with a particular type of human sweat gland: the apocrine. Apocrine glands, found chiefly under the arms and around the genitals, are inactive in childhood and begin to operate only when puberty is reached, which explains why adults have a body-odor problem and children do not. There are other sweat glands that produce perspiration for cooling the body, but these seem comparatively free from microorganisms, perhaps because of a continual flow of fluid, which keeps the pores clear. Around the apocrine glands, however, bacteria grow in large numbers. They live on substances in the sweat, and it is their digestive activity that causes the odor. Sweat collected

from apocrine glands that are free of bacteria has been found to be odorless, but it begins to smell when bacteria from the skin are introduced into it. Deodorants are really disinfectants that kill the bacteria that live on stale sweat.

Although appreciable populations of micro-organisms can thrive only in moist areas of the skin, some of the invisible creatures do inhabit the entire skin surface, especially in and around the hair follicles (the fine sheaths in which hair shafts are embedded). These channels in the skin provide a first-rate environment for microbial growth. Each hair follicle has a small gland that produces an oily lubricant for the hair shaft. Yeast and other fungi, as well as bacteria, find this an attractive habitat; and large numbers of microorganisms embed themselves around the follicle, just under the skin surface, where they stay firmly attached and cannot be removed by washing. Doctors and others who have to keep their hands and arms scrupulously sterile use a variety of chemicals to remove bacteria.

How densely populated is the human skin? Scientists at the University of Pennsylvania report that in adult males the underarm bears the greatest concentration, with an average population of 15.54 million bacteria per square inch of skin. The scalp and the forehead have estimated average counts of 9.42 million and 1.29 million bacteria per square inch respectively. (Note the huge difference between the numbers on the moist, protected, hair-covered scalp and the more exposed forehead.) By contrast, comparatively few microorganisms appear to live on the back, which houses an estimated average of only 2025 per square inch.

The skin, like the stomach, produces acidic substances that kill off some of the most harmful types of bacterium. Among the few resistant groups are the staphylococci, which are generally present in colonies resembling miniature bunches of grapes. One such bacterium, which lives on the skin of about 10 per cent of all healthy people, can damage sweat glands and hair roots, particularly in adolescence. It is at least partly responsible for acne—the unsightly spots and pimples that blight so many teenagers' hopes of physical beauty (and that have therefore opened up another profitable market for the cosmetic and pharmaceutical industries).

Although vast numbers of bacteria also live in the nose, throat, and nasal passages, very few are able to make their way down to the lungs. The

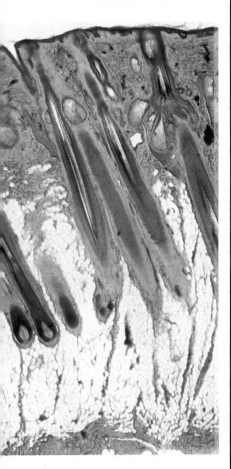

Certain areas of the skin provide favorite homes for microorganisms. Bacteria and fungi are especially numerous just below the surface in the kinds of site revealed by the skin section above. There, microbes find safe niches in the protected shafts (oil-lubricated by associated glands) from which hairs grow. Many microorganisms on skin escape casual detection. But the stale sweat from underarm apocrine glands provides food for bacteria that produce unpleasant body odor. Deodorants applied as spray (right), cream, powder, or a solid stick prevent body odor by killing the bacteria or by inhibiting perspiration (in which case they are known as antiperspirants). Aluminum chloride, petrolatum, formaldehyde, and vinegar are among the chief weapons used in the modern arsenal of anti-odor toiletries.

reason for this can be found in the physical structure of the walls of the air passages that lead from the nose to the lungs. These have a coating of sticky mucus, in which most of the microbes that we take in with every breath are trapped. Moreover, the walls are lined with whiplike cilia, and these beat rhythmically in an upward direction, pushing the inhaled organisms back toward the nose and throat, where they are expelled in the saliva and nasal mucus. A few kinds of organism, however, are so small and light that they escape this sticky fate. The bacterium that causes tuberculosis is one such invader. In most individuals any such bacteria that manage to reach the lungs are immediately engulfed and destroyed by special defensive cells, the *phagocytes*. But in people with low resistance the bacteria can survive and cause tuberculosis.

One unsolved puzzle still blocks our understanding of the relationship between microorganisms and higher animals: why are microbes that seem harmless to one individual deadly to another? Why is it that some individuals can tolerate the presence of large numbers of dangerous microorganisms without succumbing to disease, whereas others capitulate at once.

So far, we have looked chiefly at microorganism–higher animal interactions in which little damage is commonly caused as a result of infection, or in which there may indeed be advantages to both parties. Now let us center our attention on the microorganisms that behave in the way that most concerns us as human beings—the ones that impose disease and death upon us. Of the five major kinds of microorganisms—the protozoans, algae, fungi, bacteria, and viruses—it is the protozoans, the bacteria, and the viruses that are the chief agents of disease in higher animals. Only about 50 species of fungi (a relatively small number) cause deadly diseases in higher animals, and the algae do not seem to be responsible for any.

Fungal diseases come in two categories: superficial and systemic. The skin is the chief target for fungi that cause superficial diseases, because these organisms are able to grow on keratin, a substance found in skin, hair, and nails. The fungi cause itching and reddening of the skin. Ringworm of the scalp in children and athlete's foot in adults are typical superficial fungal ailments—not serious diseases, to be sure, but treatment is generally prolonged and not always successful, because very few chemicals are effective against fungal infections. Systemic fungal disease often has symptoms resembling a mild cold. The fungi that cause the disease may eventually become distributed throughout the body; this results in a general weakening, but such disease is rarely deadly.

Protozoans, on the other hand, are at the root of a number of serious and possibly fatal ailments—for example, malaria, which is caused by any one of four species of the genus *Plasmodium*. These creatures have a complex life cycle, several stages of which take place in man, the others in mosquitoes of the genus *Anopheles*. When a person is bitten by an infected mosquito, small cells called *sporozoites* are injected into the bloodstream. These, the protozoan's infective units, are carried through the body until they reach and enter the cells of the liver. Here each sporozoite sets out to reproduce itself asexually by splitting in half. The resultant small cells, called *merozoites,* reenter the bloodstream, where they can infect the red blood corpuscles of the host, who becomes very ill as a result. The merozoites produce more of their kind within the red blood corpuscles, just as the sporozoites reproduce by division in the liver cells.

Not all the merozoites released from the red blood corpuscles are capable of infecting other corpuscles; some are capable of infecting only a mosquito. When an anopheles mosquito bites an infected man, it draws up these *gametocytes* (as the infective cells are called at this stage) with its ration of blood, and so they enter the mosquito's body. The mosquito may then bite another infected person, drawing up more gametocytes—this time, of course, from a completely different parent *Plasmodium*. Gametocytes from the two parents may then fuse, producing new creatures, which, by means of amoeboid movement, make their way to the mosquito's intestine. In the mosquito's gut, the amoebalike creatures grow and divide into a number of sporozoites. These in turn find their way to the salivary glands, where they become concentrated, available for transmission to a human being.

So the best way to prevent malaria is to wipe out the anopheles mosquito. Failing that, certain drugs such as quinine, paludrine, chloroquine, and atebrine are effective measures against the protozoan during its most virulent stage, when it is in the blood corpuscles.

Plasmodium and other protozoan parasites on higher animals may seem to us to lead rather

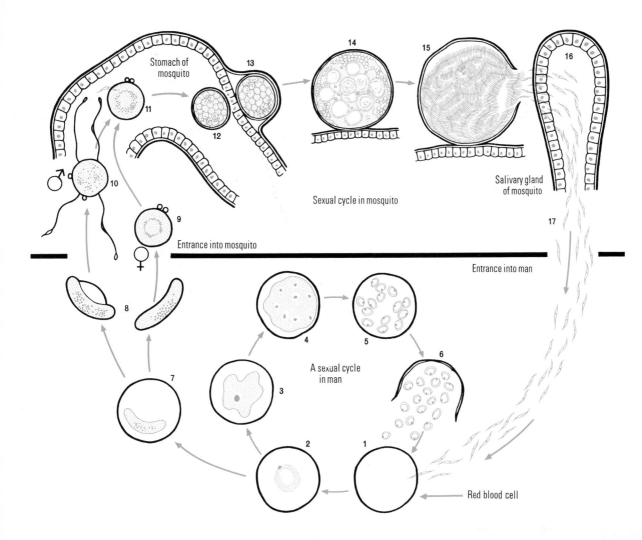

Stomach of mosquito

11

13

12

14

15

16

Sexual cycle in mosquito

Salivary gland of mosquito

10

9

Entrance into mosquito

17

Entrance into man

8

4

5

6

A sexual cycle in man

3

2

1

7

Red blood cell

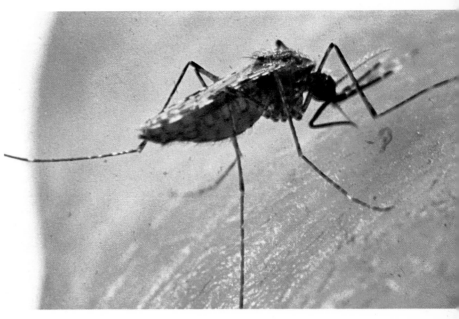

Anopheles *mosquito sucking blood from a human finger (right) also injects saliva that may contain* Plasmodium *protozoans—microorganisms that cause one of the world's chief debilitating and killing diseases: malaria. Mosquito and man play complementary roles in* Plasmodium's *complicated life cycle. Above:* Plasmodium *sporozoites entering man infect and destroy red blood cells and reproduce asexually (1–6), producing small cells called merozoites that give rise to sex cells called gametocytes (7–8). Sucked up with human blood into a mosquito's stomach, each female gametocyte becomes a mature macrogamete (9). Each male gametocyte produces microgametes (10). If one enters a macrogamete (11) the resulting oökinete (12) lodges in the outer stomach wall (13), becoming an oöcyst (14) that bursts, releasing sporozoites (15). Many reach the mosquito's salivary gland (16). These sporozoites can enter man (17).*

strange lives. But of all the weird life styles adopted by parasitic microorganisms, none is stranger than that of the viruses, as we have seen in our discussion of bacteriophages in Chapter 4. Although we have known about viruses since 1898, when two German scientists suggested that these tiny particles were responsible for foot-and-mouth disease in cattle, doctors have not yet learned how to control them. And so, because all viruses live as parasites in living creatures, it is just as well that most of them seem to coexist fairly harmlessly with their hosts. Why this should be so is not at all clear, but the fact remains that viral attack resulting in serious damage to the host is the exception rather than the rule. Furthermore, viruses that do great harm in one organism seem quite innocent in others. The encephalitic viruses, for example, are deadly to human beings but seem to live naturally and harmlessly in fowl.

Viruses have no means of locomotion and are transferred from one host to another only by wind, water, or personal contact. Similarly, they have no means of getting inside the host, but must depend on breaks in the skin (or in the roots, stems, or leaves of plants) or must enter through the mouth or lungs. But we really have little knowledge of precisely how a virus enters the body, whether of an animal or of a plant. And we have only a general idea of how it gets about once it is inside.

The target for the invading virus is an individual cell. Entry to a living cell, with its cell membrane, might seem an impossible task for a virus particle, which has neither teeth nor any other built-in tool for effecting an entrance (with the exception of the phages that parasitize bacteria). But the virus has an unusually subtle method of unlocking cell doors without breaking through the walls. Somehow it deceives the cell into taking it in as if it were food. How it manages the deception is not entirely clear, but entry evidently depends on the structure of both the virus and the cell membrane. If the viral key fits, the door opens and in goes the virus.

Once inside the cell, it takes over the cell's machinery for the construction of growth materials and redirects it into making new viral particles. Eventually, as for the phages, the newly created virus particles are released by the breakdown or bursting of the host cell. Such destruction on a large scale causes disease.

Viral diseases that afflict man include polio, encephalitis, influenza, and—most familiar of all ailments—the common cold. Because we have few ways of dealing with such diseases once infection has occurred, prevention rather than cure is the watchword. Immunization (the practice of using vaccine against viral attacks) is one such defense. Vaccines create artificial immunity to disease. When animals are attacked by hostile virus particles or bacterial cells, they produce antibodies whose function is to destroy the invaders. The process of immunization is based on the fact that once the antibodies for some diseases have been called forth, they circulate in the body and confer immunity against those diseases for quite a long time. So if a person is vaccinated—that is, artificially infected with a very mild dose of a disease—he builds up defenses against serious attacks.

Credit for discovering the value of vaccination is commonly given to an English doctor, Edward Jenner (1749–1823), who, although he did not really discover the process, found a way to make it safe. During a terrible epidemic of smallpox toward the end of the 18th century, Jenner noticed that farmers and milkmaids who had contracted the much less serious disease of cowpox seemed safe from smallpox, while those around them were dying in droves from the disease (which probably killed about 60 million people in various parts of the world during that century). Jenner wondered what would happen if he put some of the pus from a cowpox spot into

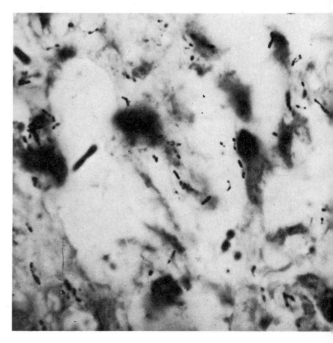

Hands reduced to stumps help to identify this Pakistani woman as a victim of the terrible and mildly infectious disease leprosy, caused by bacteria like those in the photomicrograph (left). Probably entering the body through breaks in the skin, rodlike Mycobacterium organisms multiply in skin and nerves and, if left untreated, sometimes cause bone damage that results in the loss of toes and fingers. Medicine is at last combating this disease, which affects millions of people in the tropics.

111

Edward Jenner's fight against smallpox.

Edward Jenner, the son of a clergyman, was born in 1749 at Berkeley, Gloucestershire. Although his family was not poor, Jenner became a surgeon's apprentice at the age of 13, because that was the best way to learn to be a doctor in those days.

a cut made on the arm of a healthy child; he tried the experiment, and the boy caught cowpox. Later on, in a clinching experiment, he inoculated the same boy with pus from a smallpox spot. After a few anxious days, it became clear that the boy had achieved immunity from smallpox. What had happened was that the cowpox virus, which is very similar to the smallpox virus, had called forth antibodies that provided protection against both diseases.

These days, the virus particles used for immunization against diseases such as smallpox, poliomyelitis, and measles are cultured especially to be effective at causing the body to produce antibodies but without causing the disease proper. These so-called *attenuated* particles are effective because it is the specific chemical content of the virus, not its ability to reproduce, that is the important factor in calling forth the antibodies.

If immunization is not used, or if it fails, there is little that doctors can do about a virus disease except let it run its course. Such antibiotics as penicillin, which are remarkably effective against bacteria, are not much use for controlling viruses; at best, they can merely help to moderate a viral attack. Fortunately, the body produces not only specific antibodies against specific viruses, but also a more general weapon in the form of defense substances called *interferons*. These are called forth whenever the cells are invaded by viruses, and they stop the viral particles from multiplying. Interferons are different from antibodies in that they do not combat a specific invader. Once initiated, they prevent the multiplication of any virus that attacks.

Like the viruses, bacteria must gain access to their victims without the help of teeth, claws, or any other organs specialized for breaking and entering, and they too are spread from one individual to another either through the air, by personal contact, or through infected food or drink. They do not always survive the trip, for if they cannot live on the skin surface, they will die unless they can make their way inside through a body orifice or a break in the skin. Once inside, however, they grow and multiply, frequently close to the point of entrance. If they get into the bloodstream, they may be distributed throughout the body and begin to grow in all the tissues that they reach, although it is more usual for a particular bacterium to grow only in a specific tissue.

Brucella abortus, a spherical bacterium that causes contagious abortion in cattle, concentrates in the unborn offspring and maternal fluids of the pregnant cow, and abortion results in about four out of every 10 cases. This same bacterium can also infect human beings, but without injuring the fetus; instead it causes undulant fever, characterized by chills and a fever that increases at night and drops during the day. Apparently the embryonic tissues of a cow contain a chemical that stimulates the growth of these bacteria. But people do not contain this substance, and so the bacteria becomes widely distributed instead of concentrating on a specific tissue.

The most specialized form of bacterial transmission through bodily contact is found in the highly contagious venereal diseases, gonorrhea and syphilis. Syphilis, the more virulent of these two social scourges, is caused by an unusual type of bacterium, one of the spirochete family. The spirochetes are slender, spiral-shaped cells, which may be of quite considerable length. They move with a snakelike motion as the spiral cell alternately contracts and relaxes—a unique form of locomotion in the microbial world. Gonorrhea bacteria cause local inflammation of the urinary and genital organs, but rarely produce serious complications or death. Syphilis bacteria, on the other hand, may spread from these sites to other parts of the body and may cause damage to the central nervous system, producing paralysis and insanity. Both these types of bacterium, however, are easily killed by adverse conditions. They cannot survive desiccation, extreme changes of temperature, or strong ultraviolet light. In fact, they are so sensitive to heat that, before the advent of antibiotics, doctors used to treat syphilis by raising the temperature of the patient's body—perhaps by inducing a fever artificially—in order to kill off the spirochetes.

Various other bacteria damage their unwilling hosts in a number of ways. Sometimes—though rarely—they do great harm just by sheer weight of numbers. A large mass of cells can block veins, arteries, or the valves of the heart, or they can clog the air passages in the lungs. But they are more likely to bring on diseases through the production of poisonous substances. The bacterium that causes diphtheria, a club-shaped microorganism that lives in the respiratory tract, is of historical interest because it was the first bacterium to be recognized by scientists as a

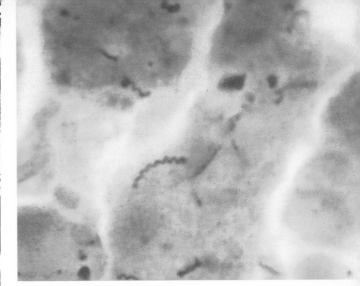

Courtship (below) may lead to sexual union in which a partner with venereal disease can unwittingly transmit the disease bacteria to the other partner. The spirochete bacterium Treponema pallidum (photomicrograph, above) causes syphilis, worst of all venereal diseases. Old "cures" included taking mercury (woodcut, left), but penicillin is the modern treatment.

poison-producer. The poison is an excretory product, and it can be as unfortunate for the bacterium as for the victim, because the microbe is unlikely to survive the death of its host.

Bacterial poisons can damage virtually any part of the body. Some, for example, attack the nervous system. One such poison causes tetanus by attaching itself to the point where one nerve fiber makes contact with another. The result is that the message being passed between nerves is hindered or canceled, with consequent paralysis in the affected muscles.

All the many instances of how man and other animals can be attacked, invaded, and in some cases killed by parasitic microorganisms might lead you to believe that the higher animals are sitting targets for microbial disaster. Fortunately, this is far from true. In addition to production of antibodies and interferons—or, as biologists call these defense mechanisms, "the immune response"—at least two powerful natural defense systems are always at work in our bodies. The first is the presence of *phagocytes*—a name meaning literally "cells that eat." And this is just what these cells, which resemble amoebas, do: they seek out, engulf, and swallow menacing alien cells. They reside in the bone marrow and in the bloodstream, where they are popularly known as "white corpuscles," and their numbers increase enormously during an infection. When a bacterium has been engulfed by a phagocyte, the cell's digestive juices swiftly destroy the microbe. It is interesting to note that practice makes perfect with these hunter cells. Once a phagocyte has swallowed its first invader, it becomes about 10 times more effective at tackling the next alien particle of the same type.

This is a very effective scavenging system. From 80 to 90 per cent of all foreign particles that enter the liver in the blood, for example, are removed by the phagocytes at one sweep. In the end, though, aggressive bacteria occasionally overcome these steadfast warriors.

The second line of defense is inflammation. Increased blood flow to an affected region of the body causes the tissues to swell, with accompanying heat and pain. The swollen tissues are likely to produce blood clots, which can then trap the invading bacteria in their meshes, making it easier for phagocytes to pick off the invaders.

As we all know, however, no defense is foolproof. The immune response is the most specific—and in many ways the most effective—protection against microbial invasion; yet the amazing flexibility of microorganisms enables them sometimes to circumvent even this obstacle. Some bacteria, for example, coat themselves in an outer layer of material that resembles the victim's own proteins. Because the body releases antibodies only when threatened by alien proteins, it does not react against such a camouflaged intruder, and so the bacterium makes its entrance under false colors. For example, *Streptococcus pyogenes*—which can cause a number of diseases in man, ranging from a sore throat to pneumonia and rheumatic fever—often escapes pursuit because it is coated in a chemical common in human connective tissue.

Not all diseases are caused by microorganisms; some are inherited, and others may come as a result of poor diet, a rigorous environment, nervous strain, or a number of other disabilities. And the microorganism is only one factor in the disease with which it is associated. We have seen something of the balance between the weapons of the invader and the defenses of the host. The microorganism may be a necessary cause, but it is not sufficient in itself. For infection to take place, the host must be susceptible, and not all hosts are equally sensitive to microbial attack. The health of the host, the vigor of the microorganism, and the presence or absence of specific immunity all contribute to the outcome of infection. In an infectious disease, potential victim and microorganism interact.

It is easy to believe that because some of the invisible creatures cause disease, all of them are harmful to man. Not only is this not true, but it is actually *very* wide of the mark. Most microorganisms are either harmless or beneficial to man, whether directly or indirectly. Much of the world's food depends on the nitrogen-fixing activities of microorganisms, and their activities in the digestive systems of cows and sheep play a massive part in helping to nourish and clothe us. As we shall see in the next chapter, the tiny creatures also confer other benefits upon us. And so, even while acknowledging the damage that some microbes do, we should not forget that the other side of the picture is very bright indeed.

In this photomicrograph an amoebalike macrophage (a phagocytic cell) is engulfing a carbon particle that has accidentally entered a lung. Macrophages acting as police and street cleaners are part of the body's first line of defense against invading, hostile microorganisms and other foreign bodies.

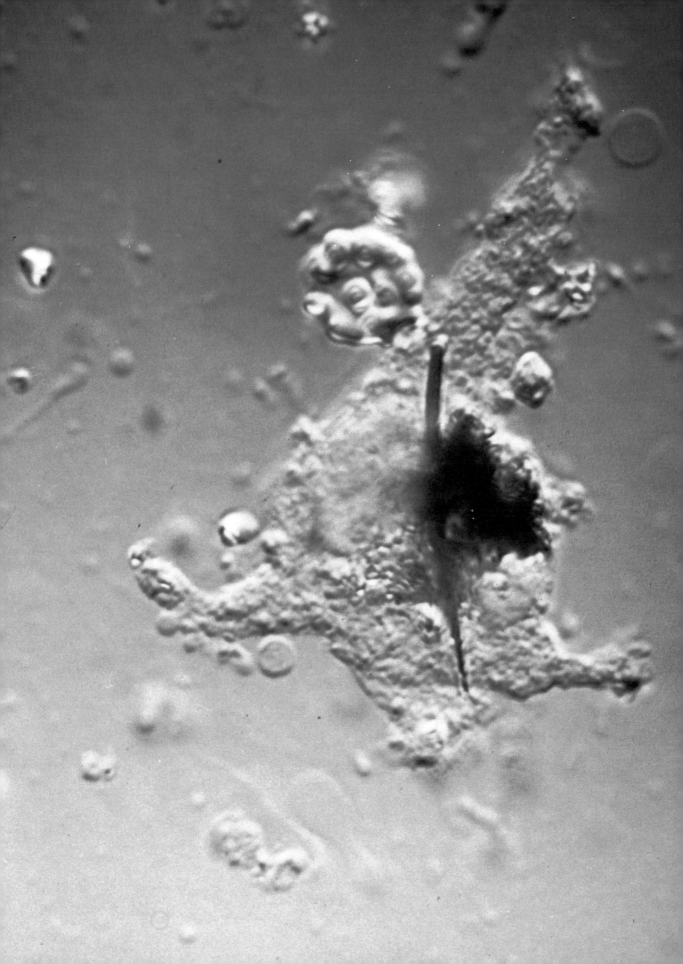

Destroyers and Creators

Biodeterioration and *biodegradation* are the fearsome names that modern scientists give to two aspects of microbial life that man has been familiar with ever since building his first primitive shelter and tasting his first alcoholic drink. *Biodeterioration* is a term that covers all the many different ways in which microorganisms (and some larger creatures as well) are able to attack and damage materials, whether mountains or masonry, fish cakes or fuel lines. *Biodegradation* includes all the processes through which living creatures break down waste matter, or convert complex substances into simpler ones.

As so often where microorganisms are concerned, there are both benefits and disadvantages for man in these activities. On the one hand, the overall cost of preventing microbial damage to food and fabric, and of repairing the damage where prevention has failed, is astronomical. On the other hand, without the unwitting help of microorganisms there would be no cheese, yogurt, beer, or wine, and the disposal of waste and refuse would be an even greater problem than it is. Moreover, many of the minute creatures whose insatiable appetites can destroy living human flesh are themselves susceptible to attack by other members of their invisible world. And so, in the use of antibiotics, for example, we have learned how to use their capacity for doing damage to one another as a weapon against those that do damage to us.

Antibiotics, which are chemical substances produced naturally by some fungi and other microorganisms, were discovered by the Scottish bacteriologist Sir Alexander Fleming less than half a century ago. In 1928, while growing cells of the bacterium *Staphylococcus* in his laboratory, Fleming noticed that one of his preparations had become contaminated by a mold-producing

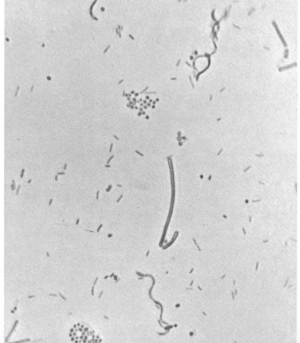

Dots and squiggles (right) are wall-grown bacteria, much enlarged. Such tiny, harmless-seeming organisms flourishing on stone in damp air produce acids that can help rot a city. Eroded statue of an angel (above) and crumbling canal-front buildings (far right) reveal the decay that threatens Venice.

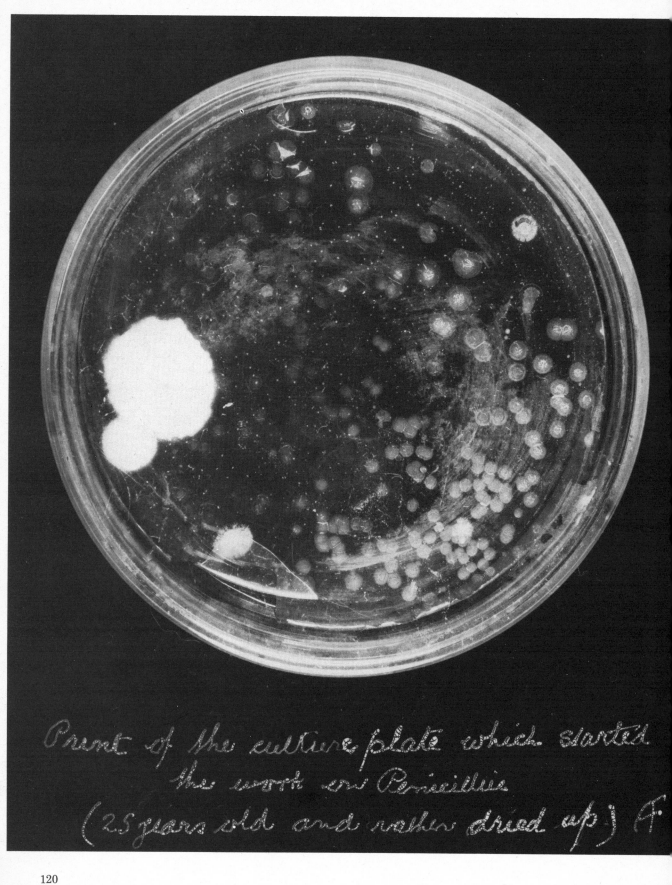

Print of the culture plate which started
the work on Penicillin
(25 years old and rather dried up) F

fungus. He also observed that the area around the fungus was entirely free of living bacteria. From this observation and subsequent experiments, Fleming discovered that the intrusive fungus was manufacturing a substance that prevented the growth of the bacteria. Because the fungus was identifiable as *Penicillium,* he gave the name *penicillin* to the substance.

Penicillin was only the first of many antibiotics. We know very little about how they work when introduced into diseased tissue, but we do know that these microorganism-produced substances have a lethal effect upon other microorganisms (chiefly, though not exclusively, bacteria). Some of them seem to prevent bacteria from manufacturing their cell walls; others kill the bacterial cell membrane; still others inhibit bacterial protein synthesis. We understand almost nothing about the role of antibiotics in nature. Some scientists believe that they are actively involved in the competition among rival microbes for living space and food—but it also seems possible that only microorganisms grown in the laboratory produce such boons to modern medicine as penicillin and streptomycin. At any rate, they *are* boons. And they compensate, at least partly, for the havoc wrought by other forms of microbial life.

Because of the apparently endless variety of ways in which microorganisms get the energy that they need for life, they are able to use—in other words, to damage—virtually anything. Most living creatures get their energy only from the breakdown of sugars, protein, and fats, but microbes can survive on such foodstuffs as metal, paper, or paint. There are bacteria able to live on iron, and they can reduce iron and steel pipes to crumbling shells. And there are bacteria and fungi that eat the rubber insulating layer around the metal core of electric cables. So great is this problem that where a cable is to be laid in a particularly exposed position, a fungicide—a chemical that kills fungi—may be incorporated into the rubber of the sheath.

What might seem, at first glance, a bizarre example of the way microorganisms are able to feed on unlikely materials is to be found in the huge tanks used for bulk storage of fuels, especially petroleum and kerosine. When these tanks are filled, it is impossible to prevent some moisture from entering with the fuel. This eventually separates out as a layer of water under the fuel (for water is the denser of the two liquids), and bacteria and fungi grow in great numbers where the fuel and water are in contact. This is, in fact, not surprising. Petroleum is a rich source for the kinds of chemical on which these microorganisms live; after all, the production of petroleum was partly the work of long-dead microorganisms, and everyone knows that petroleum is a source of energy.

Oil-eating microbes have become a severe hazard in jet-airplane fuel systems, which burn kerosine. If microorganisms get into the kerosine, the fuel filters become clogged with microbial debris, with resultant power loss; furthermore, microbes growing in an aircraft's fuel tanks can cause corrosion. To minimize such dangers, the kerosine is strained through special filters, and chemicals that prevent the growth of bacteria and fungi are added to it. In addition, the inner walls of the fuel tanks are coated with a plastic substance that is resistant to attack by microorganisms, and the tanks are washed out at regular intervals with a lethal substance.

It is not surprising that microorganisms devour wood, but it may be surprising to learn that they also damage stone. Indeed, in attacking the surface of exposed rock faces and boulders, they often break off tiny particles, which constitute the first stage in soil formation. They also threaten the structure of stone buildings and contribute to the gradual wearing down of old statues and carvings.

How do such miniscule creatures manage to damage great rocks? If you examine almost any large area of exposed stone, you will find algae and lichens either lying dormant in dry weather or growing profusely in wet weather. The foods that they produce through photosynthesis also support the growth of other microorganisms, such as bacteria and fungi. All these single-celled organisms breathe in oxygen and give out carbon dioxide gas, and as the carbon dioxide dissolves in the surrounding moisture, it forms carbonic acid. This, together with other acids produced naturally by the microorganisms, helps to dissolve the surface of the rock, causing small particles to break away.

The culture plate on which Fleming first noticed Penicillium *mold spreading and curbing the growth of* Staphylococcus *bacteria. The big double blob is the* Penicillium. *Smaller blobs are* Staphylococcus *colonies. Between the two is an area where penicillin produced by the mold has caused the bacteria colonies to begin to break up. (This photograph was taken when the culture was 25 years old and rather dried up.)*

Meanwhile, the cracks and fissures in the rock that result from freezing and thawing gradually become filled with soil, which allows the growth not merely of larger plants but of further colonies of microbes. And so the deterioration caused by the combined efforts of weather, microorganisms, and larger plants progresses.

Even road surfaces deteriorate because of bacterial growth on substances present in asphalt and bitumen. And among other materials that you might wrongly assume to be impervious to attack is the glass of lenses and mirrors. The glass itself is not eaten by microorganisms. What happens is that tiny amounts of it are damaged by the acids that they produce, but the minor destruction is enough to spoil delicate optical surfaces. In the tropics especially, certain types of fungus grow on the dust around lenses or on the adhesives used in mounting the lenses in cameras and binoculars, or on the special substances that coat lens surfaces. The fungal acids etch the glass, and even the use of fungicides and meticulous efforts to keep surfaces dry and dust-free will not always protect them from invasion.

Paints, too, are subject to various forms of biodegradation. Fungi growing on the timber to which paints are applied may discolor them, or they can be damaged by fungi in the dust on the paint surface. There is a quite common bacterium that goes to work on paint as it is being used, and other kinds of bacteria often thrive inside stored tins of paint.

In fact, there seems to be only one kind of material substance that comes close to defeating the microbes: man-made plastics. There are some bacteria that attack plastics, including polyethylene—but it is slow going for them, and they do not get far. As we have seen, microorganisms can break down virtually every other substance, returning the chemicals from which it is made to the soil, sea, or air. But plastics are a remarkable exception to the rule. It has been suggested that as archaeologists of the future dig their way down through the layers of debris left by successive civilizations, they will be able to recognize ours instantly because our remains will be composed chiefly of discarded plastic wrappers, bottles, cups, bags, and other such indestructible items.

The reason why the microbes are so baffled by plastics is that they must, like any other form of life, obey the rules of chemistry. They can degrade materials only if they are able to produce chemical substances that can act upon the material to be degraded. No microorganism had ever confronted a man-made plastic before the early part of this century, and so none had ever evolved chemicals to deal with such an unnatural material. No doubt the inexorable force of the evolutionary process will eventually produce a microbe that can change a plastic bag into a fine dinner. In the meantime, modern science is doing what it can to help solve the problem.

For it *is* a problem, and an oddly paradoxical one: while science searches for new and better ways to protect materials against microbial attack, science also searches for ways to make one kind of material vulnerable to the microbes. Unless either natural evolution or science comes to our rescue, the modern world is in danger of burying itself under a mountain of non-biodegradable plastics!

The scientists' efforts have recently met with some success. The British, for example, are working on the development of a plastic that would disintegrate when exposed to direct sunlight. Polyethylene, polypropylene, and other commercial plastics could then be manufactured in such a way that wrappers and containers made from such substances would remain usable in diffuse daylight, and even in sunlight filtered through a window pane; but exposure to direct sunlight would initiate a process of disintegration leading to the formation of a harmless granular substance, which would be composed of chemicals that bacteria could recognize and break down. By adjusting the amounts of the various constituents of the new plastics, its makers could probably vary its outdoor lifetime from about three months to five years.

Apart from plastics, however, practically all materials provide both habitats and food for the invisible marauders. And some of the destruction that they wreak on substances of value to man are actually tragic. Among the most serious kinds of microbiological deterioration—to us, at any rate—is the spoilage of foodstuffs. What this costs economically is impossible to estimate accurately, but the global figure must be billions of dollars annually.

We describe food as "spoiled" when it is no longer attractive or safe to eat. The loss is due not to the amount of food that the microorganisms take (although, given time, they will

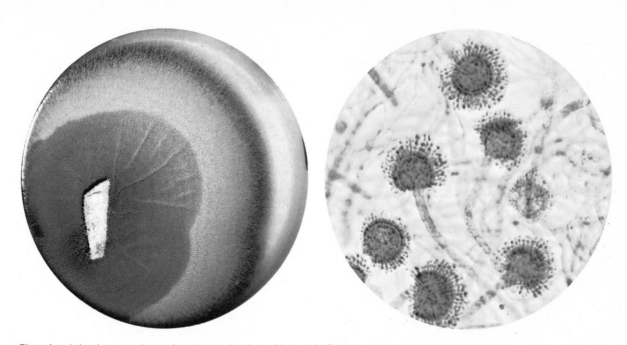

Three fungi that harm products of modern technology. Above left: Cladosporium resinae *culture growing on an aircraft filter. This fungus corrodes the interiors of water-contaminated metal fuel tanks. Above right: photomicrograph of an* Aspergillus *fungus. Members of this genus erode optical lenses. Below: here a frothy mold feasts on insulation material sheathing an electric cable.*

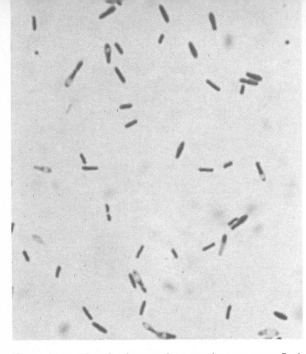

completely consume a whole carcass or an entire field of grain) but to the substances they produce while feeding.

These chemical by-products of the feeding process may render the food completely unfit for human consumption simply by making it taste or smell foul. In many cases the smell and taste are all that makes the food unattractive, for it is still perfectly safe to eat. In other cases, however, microorganisms produce poisonous substances that can cause serious or even fatal diseases. A classical example of such a disease is the bacteria-generated botulin poisoning that we discussed in an earlier chapter.

Microorganisms are present in such large numbers in the air, soil, and water, as well as on the skin and in the bodies of man and other mammals, that it is extremely difficult to prevent food from becoming contaminated, especially if it has to be kept for any length of time before being eaten. For example, the flesh of cattle butchered for beef probably contains very few microorganisms; but the skin and intestines are sure to be highly contaminated, and the

Germs rot or poison food exposed at normal temperatures. Rod-shaped Clostridium botulinum *(above, enlarged 500 times) breeds in badly canned food and the food can kill if swallowed. But efficient canning prevents bacterial growth. Food abandoned in Antarctica by Scott's last expedition (photographed below and right) thus proved sound after years in the Southern Hemisphere's great polar icebox.*

contaminating organisms will in turn contaminate workers in abattoirs. The workers and their instruments will then pass the microbes on to the flesh. Furthermore, if an animal is inefficiently killed, its heart may continue to beat for a time. In such an event, any organisms that manage to gain access to the bloodstream through open wounds will be swept around the body as the blood circulates.

Fortunately, the standard of cleanliness in most modern abattoirs is high, and freshly slaughtered flesh is quickly chilled to prevent the growth of all but a few bacteria. Cooking provides the final deathblow for these bacteria, which cannot survive high temperatures. Real danger arises only when meat is allowed to stand for some time at temperatures that permit the growth of microorganisms.

Food spoilage is so much a part of our common heritage that the words used to describe the different sorts of stale food are more folksy than scientific. Thus we traditionally talk of moldiness and "whiskers" when foods become covered with spots of fungus, "sliminess" when the surface of meat or fish is damaged by bacteria, and "ropiness" when bacteria cause a sticky material to form in wine, vinegar, or milk. In this century, however, some new terms have been added to the food-spoilage vocabulary because certain forms of spoilage have become possible only through the advent of modern food technology, especially canning. Although canning is a technique used specifically to preserve food from microbial attack, the technique itself involves risks. "Flat sours" is the odd name given to spoilage caused in canned foods by organisms that ferment the food without producing gases as a by-product. "Blown cans" describes what happens when another kind of bacteria ferments food inside cans, but this time *with* the production of gas: the gas distends and distorts the walls and ends of the can and may eventually blow it open.

The contamination that results in such spoilage as "flat sours" or "blown cans" happens infrequently as a consequence of some sort of error in the canning process, but it does happen, and it can sometimes have tragic or near-tragic

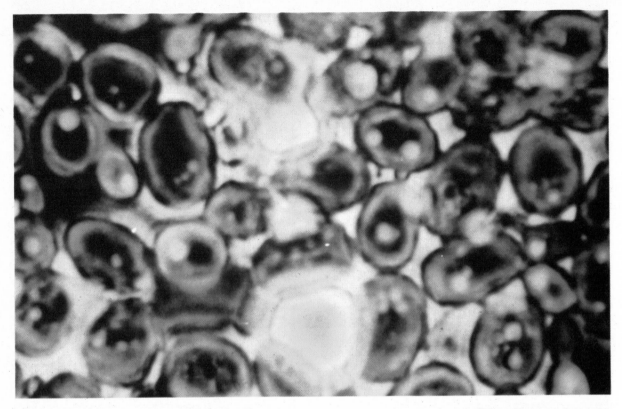

Photomicrograph of yeast cells growing on gas oil. The resulting crop is concentrated, dried, and purified, and the end product is a protein used to feed farm livestock. In such ways, micro-organisms may help combat a mounting world food shortage.

Right: salmon hanging in a Suffolk smokehouse in eastern England. Warm smoke from hardwood sawdust dries the fish and impregnates it with chemicals including wood creosote. This process confers flavor and protection against decay bacteria.

results. Some years ago, cases of typhoid broke out in areas where people had eaten canned corned beef from Argentina. The faulty cans had been cooled in Plata River water after an improper sealing, and poisonous *Salmonella* bacteria, which abound in the river, had entered the cans and multiplied in the meat.

Some foodstuffs can be contaminated with harmful microorganisms in their raw state. For instance, the eggs of hens and ducks can be infected by *Salmonella* bacteria, which have been known to attach themselves to the outside of the shell, penetrate it, and grow inside. But this sort of attack is fortunately rare, and so food scientists are chiefly concerned with the problem of maintaining the quality of meat or of plant produce in the period between the death of the animal or the gathering of the crop and its consumption. In the tropics, where warmth and humidity are conducive to the swift growth of bacteria, decay takes place very quickly indeed. This explains the traditional use of curry powders, herbs, and spices in hot countries;

such strong seasoning masks the unpleasant odor and taste of decaying but usually harmless food.

Nowadays we have a number of different ways to prevent—or at least delay—spoilage. The first is refrigeration, which, although it does not kill bacteria, slows down the rate at which they can grow and reproduce themselves. When food is kept frozen at very low temperatures it can be stored almost indefinitely, for frozen micro-organisms remain totally inactive. Supplies left in the Antarctic by the ill-fated Scott expedition in 1912 were discovered many years later in perfect condition. And mammoths that died and became embedded in the Siberian ice fields many thousands of years ago have been dis-interred with their flesh still edible.

At the other end of the temperature scale, food is also preserved by great heat. Canning and bottling, for example, work on the principle of killing bacteria already present and ensuring that no others can get in. To do this, the food is thoroughly cooked, then placed in sterilized cans or bottles and reheated under pressure, and

the containers are sealed while still very hot. This is a useful technique, but it must be strictly controlled or it can go wrong.

Pasteurization, named for its inventor, Louis Pasteur (1822–95), is a heating technique for preserving milk and other fluids. Two general methods are in use today. The traditional way of pasteurizing milk involves heating the liquid to about 160°F for 15 seconds, then cooling it quickly to about 54°F and pouring it into sterilized bottles, which are capped immediately. This is called "flash pasteurization." In a more recent modification of this technique, the fluid is held at 150°F for 30 minutes before being cooled. But an entirely different—and much newer—method of preserving milk is to heat the liquid to an extremely high temperature— about 270°F—for one second and then instantly seal it into containers. This is how modern long-lasting milk is treated so as to remain "fresh" for months. The very high temperature kills many more of the contaminating microorganisms than does traditional pasteurization, though it does cause the milk to have a different flavor.

Food may also be preserved by drying it or by immersing it in high concentrations of sugar or salt. Bacteria cannot live without moisture, and in highly concentrated solutions the water is drawn out of them. More modern techniques involve the use of such chemicals as sulfur dioxide, proprionic acid, and sodium nitrate. These are *bacteriostatic*—that is, they prevent microorganisms from multiplying in food, but do not kill them. Very small amounts of these preservatives are used, because large doses of them can be dangerous to human beings.

Pickling food in vinegar and the smoking of meats and fish are also examples of chemical preservation, although most of the people who have been pickling and smoking their foods for many generations have hardly been aware of that fact. Vinegar, of course, is a weak acid, and acids either kill bacteria or control their growth. As for smoking, the special flavour that it adds to certain kinds of food is only one of its virtues; the other is that the smoke impregnates the surface of the food with microbe-inhibiting chemicals. Combined with the heat of the process, the effects of the smoke enormously reduce the number of bacteria in a piece of fish or meat.

To pass from the most old-fashioned to the most newfangled of preservation techniques is to move into our age of controlled radio-activity. Because bacteria are killed by ultraviolet light, which is a natural constituent of sunlight, we now sterilize all sorts of things by means of radiation. You may have noticed that electric razors for general use in airports and other public places are often enclosed in boxes filled with blue light. This is ultraviolet light, used to kill "germs" that might otherwise be transferred from one person to another. Ultraviolet light cannot penetrate very far into foods, though. So we use a much stronger kind of radiation as a preservative. The food is first sealed into a container, and then exposed to radiation from a radioactive cobalt source. Because the cobalt never comes near it, the food itself is not contaminated with radioactivity, but the radiation that passes through the packaging and the food kills all microorganisms. Any edible matter treated in this fashion really does become completely sterile.

Some foods, however, are *improved* by being infected. Best steak, for example, benefits from "hanging" for a certain time, because the by-product of bacterial activity is a steak that is more tender and tasty. It must not hang too long, though, or the good taste will turn bad. But the wide range of natural dairy products is a familiar example of good-to-eat infected food.

Untreated raw milk quickly goes "bad," of course. This is because it is attacked by various species of bacterium that convert milk sugar into an acid that sours it. Yeasts and other fungi can live on this acid. Then, when they have used it up, bacteria—whose growth has so far been inhibited by the acid—take over and degrade the milk proteins, while still other microorganisms degrade the milk fats. And so the soured milk turns rancid—unless some or all of these processes of biodegradation are taken advantage of for the production of such foods as butter, cheese, yogurt, buttermilk, sour cream, acidophilus milk, Icelandic *skyr,* and exotic intoxicating beverages such as kefir and koumiss, which come from southern Russia.

The dairy industry today is big business, and old-style hit-or-miss methods of producing milk products have given way to precise techniques, which make such production possible on a large scale. Butter, for example, is made from milk fat—that is, the cream—acted upon by two different sorts of organism. Nowadays, after the cream has been pasteurized, a special "starter culture" of the right bacteria is used in order

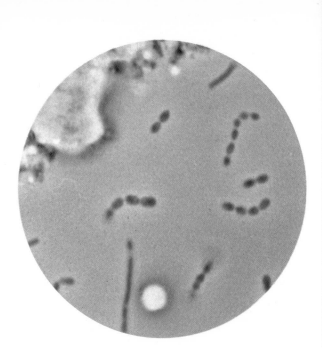

Few people buying their gaily colored tubs of fruit-flavored yogurt (right) know that this tasty beverage is manufactured by millions of tiny bacteria like these (photomicrograph, above).

to get the process going. In the old days, however, when time and mass production were less demanding than they are now, microbes floating in naturally from the atmosphere would have done the job.

Yogurt is started in a rather similar way. Even on a very small scale, it pays to use a starter culture; and, in fact, a little of the preceding batch of yogurt is often used as such a culture for each successive batch. The starter is added to pasteurized milk. When this is incubated at about 95°F, the microorganisms produce acids by fermenting the milk sugar. These acids coagulate proteins in the milk, and there you have it: yogurt. Kefir, koumiss, and other fermented milk products are all produced in basically the same way, but tastes and the available raw materials vary.

The end-product—whether liquid or solid, fizzy or alcoholic, thick or thin—depends on the kind of milk, the types of microorganism that go to work on it, the permitted degree of fermentation, and so on. "Fermentation," incidentally, is a word that we use mainly when referring to the decomposition of such organic substances as sugars and other carbohydrates. The chemical breakdown of proteins by the microorganisms is called "putrefaction."

Bacteria play a major role in the development of acidity and specific flavors in cheese. A basic cheese is made, quite simply, by curdling milk with rennin, an enzyme that coagulates it. This coagulated milk—or cheese, as it has become—is stored; and during its period in storage, the flavor and aroma develop as a result of the slow growth of certain types of microorganism within it. Cheeses of the Cheddar, Cheshire, and Wensleydale varieties are said to be "bacterially" ripened, and need no further microorganisms added during the storage period. But others, such as Camembert and Brie, are encouraged to grow mold on their outer surfaces. And some—Stilton and Danish Blue, for instance—are pierced with long rods to allow mold spores to reach the inside of the cheese. Some of the invisible organisms that flourish in cheese produce gases as a result of the fermentation process; and the gases, trapped and with no outlet, forcibly open spaces within the cheese. Thus are formed the holes in such varieties of cheese as Gruyère and Emmental.

The other universally known and enjoyed product of microbial fermentation is bread, which owes its several characteristic textures and flavors to various species of *Saccharomyces*, a minute fungus commonly called *yeast*. When yeast is added to flour-and-water dough, it leavens the mixture—which means that the fungi ferment the sugars that are in the dough and thus produce carbon dioxide gas. This makes

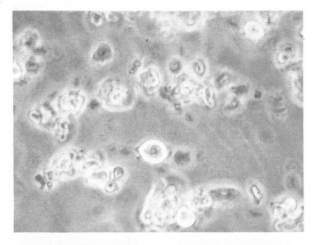

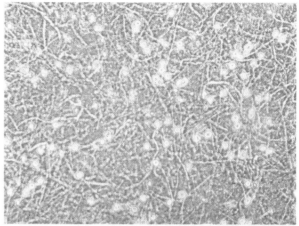

Each cheese on these pages appears below a photomicrograph revealing something of microorganisms found in it. Different microbes help to produce different cheeses. Bacteria known as streptococci *and* lactobacilli *ripen English Cheddar (above).*

Mold-ripened cheeses include Brie, a soft French cheese from the plateau east of Paris, and a close relative of Camembert. With such cheeses, a surface smear of Penicillium *mold forms a white film and softens the cheese from the outside inward.*

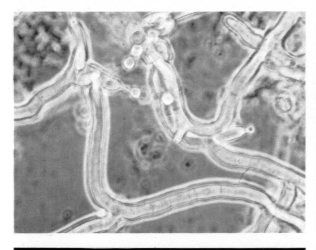

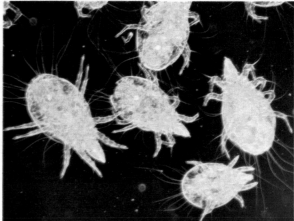

Stilton, a semihard mold-ripened cheese from Leicestershire and Derbyshire in England, gets its bluish interior veining from Penicillium roquefortii. *This mold grows in crannies inside the cheese. Molds, yeasts, and bacteria thrive on the crust.*

The same mold that gives Stilton yields semisoft Italian Gorgonzola. But not all tiny organisms in cheese are equally beneficial to us. The cheese mites in the photomicrograph above do nothing to create cheese: instead, they avidly devour it.

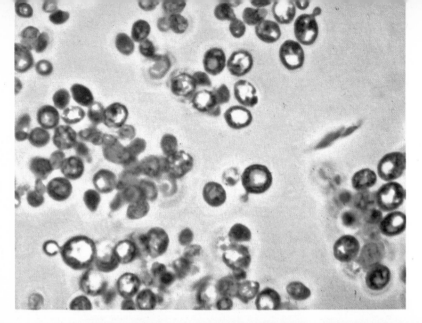

Bread-making exploits yeast's ability to breed explosively. Left: asymmetrical "dumbbells" (shown much enlarged) are Saccharomyces yeast cells multiplying. Each baby simply buds off from a single parent. The growing cells produce chemical agents that convert the starch in dough to sugar and the sugar to alcohol and carbon dioxide gas. Gas bubbles make the dough rise. Yeast breeds so fast that a mere pinch doubles the size of a lump of dough in two hours. After fermentation comes baking. This kills the yeast cells and causes alcohol to evaporate. Below: shoveling finished loaves from a baker's oven in France.

the dough less dense and increases its volume.

The fungi also add a "yeasty" flavor to the mixture. This is splendid as long as it remains within bounds, but if the taste of yeast becomes too strong, the flavor loses its delicate appeal. Thus the objective in bread-making is to permit the yeast to produce just the right amount of gas, for too many yeast cells spoil not only the taste but also the consistency of the loaf. This is why bread must be popped into the oven at just the right moment; the extreme heat of cooking kills the yeast cells.

Considering how much bread is baked and eaten in the world, it is not surprising that yeast is grown on an industrial scale these days. It is nurtured in fermentors—huge steel vessels ranging in capacity from 5000 to 15,000 gallons. A 15,000-gallon fermentor can be as high as a three-story building. The yeast cells in the fermentor are nourished on molasses and a number of other substances that cause them to multiply fast, but without producing much of the carbon dioxide gas that a baker expects when he adds them to his dough. In 12 hours inside a warm well-aerated fermentor, the number of yeast cells increases five-fold. At regular intervals some of them are removed from the fermentor and molded into blocks, which are then stored under refrigeration.

Dried yeast, which you can buy from your local store, is in reality nothing but a vast collection of spores, which will germinate to yield free-living yeast cells only if you place them in a suitable environment, such as a baking mixture or sugar solution. Because they are not adult fungi but merely spores, they remain alive despite desiccation. Yeast itself dies if kept in a dry state for a long time.

A kindred craft to baking is brewing—a craft whose products, some people would argue, are nearly as essential to man's well-being as dairy foods and bread. Wines and beers are notably various in texture and flavor; yet although scientists have studied them intensively, we do not know the exact reasons why one wine or beer is so very different from all others made by similar means. It is an interesting fact that biochemists can precisely define the quality and purity of an antibiotic produced industrially by microorganisms. And, given certain specific microorganisms grown in certain prescribed fashions, they can exactly duplicate the antibiotic under consideration. But they cannot do anything similar with wine and beer. Brewing and wine-making have always been—and will remain—more nearly in the realm of the arts than in that of the sciences.

After the grapes from which most, though by no means all, wines are made have been crushed, the unfermented juice—called *must*—is a suitable medium for the growth of yeasts and molds. Because it is an acidic solution, bacteria do not grow in it. Traditionally, the microorganisms that ferment the must—a species of *Saccharomyces*—are present on the skins of the harvested grapes, having arrived there from the atmosphere. These are known as "wild" yeasts, because they have not been cultivated or modified by man. In modern times, however, a cultivated culture of yeast is added to the must, which is then kept in large vats and well aerated while the fungus goes to work on the sugar in the juice.

Up to a point, the yeast population grows rapidly and forms alcohol and carbon dioxide gas. In general, however, growth stops when the concentration of alcohol in the must has reached about 14 to 15 per cent; after this the yeast cells are killed off by the alcohol that they have produced, and fermentation ceases. That marks the end of the microbiological part of the winemaking process and the beginning of a second stage, during which the newly fermented wine is stored and permitted to age. Now it is at risk from any other stray microbes, which could easily spoil the flavor and quality of the maturing wine. It was Louis Pasteur who first examined this problem scientifically and made suggestions for overcoming it. What we now call pasteurization was again his chief recommendation, for he pointed out that if the wine is heated to 140°F for a few minutes after fermentation has been completed, alien microbes are killed without damaging the wine. He also emphasized that the use of pure yeast, free from bacterial contamination, and the maintenance of scrupulous cleanliness in all the wineries and breweries would go a very long way toward eliminating diseased wine and beer.

Beers are made from barley, which contains starch, a substance that yeasts cannot break down. Thus the first step in beer-making is to convert the starch to sugar. Barley will do this on its own account if left to germinate, for the conversion of stored starch to sugar is one of the first stages in the germination of this seed. So barley grains are allowed to germinate for a

little while. Then, when some of the starch has been converted to sugar, the barley is dried, and the dried, partly germinated barley is what we call *malt*. The next step is to grind the malt in water, boil it, and filter it. This results in a liquid that brewers call *wort,* which is next mixed with hops, to give the beer its characteristically bitter taste.

The brew is now ready for fermentation, which is started by the addition of a special kind of yeast belonging to a group called "beer yeasts." These fungi do not live free in nature, as do the wine yeasts, but have been bred by man through the course of centuries. Often spoken of as "tame" yeasts, they resemble their "wild" cousins about as much as an Alsatian dog resembles a Siberian wolf. A brewer's success has always been measured by his ability to cultivate a yeast that makes good beer and to transfer it from one brew to the next without contamination by wild undesirable microorganisms. Special types of tame yeast have therefore been developed as a result of many long years of breeding, according to basic principles that differ very little from those governing the breeding of horses or roses.

Through bacterial activity, beer and wine (and, indeed, other beverages, such as cider) can undergo a further transformation, which also results in a product useful to man: vinegar. The organisms responsible are a small group of bacteria, the acetic acid bacteria. What most gourmets consider the best wine vinegar is produced by the so-called "Orleans process," named after the French town of Orleans, where it originated. In this process, wooden casks are half-filled with wine and placed on their sides, with holes bored in the uppermost part of the cask to let the air in. With the air come the bacteria, which float into the cask and form a thin layer on the surface. The bacteria get the energy that they need for growth and reproduction by converting the wine alcohol into the dilute solution of acetic acid that we call *vinegar* (from the French words *vin aigre,* meaning "sour wine").

In all the processes that we have mentioned so far, invisible organisms actually build up

The popular unfermented juice from Concord grapes (left) retains a fresh-grape flavor. But juice exposed to yeasts ferments and changes into wine. Right: Madeiran peasants using an ancient stone press to squeeze the juice from grapes. Above: frothy surface of fermenting juice that will become port wine.

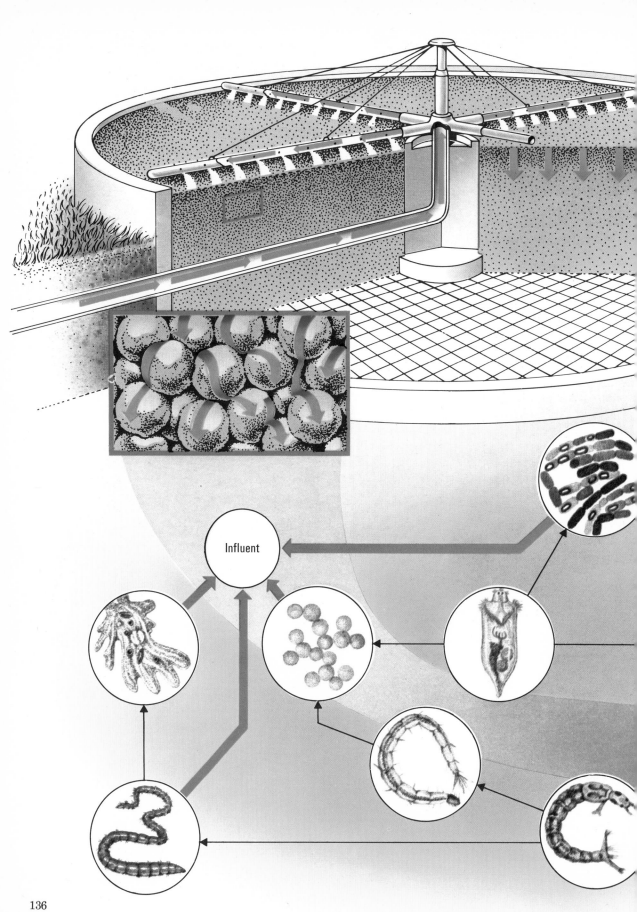

Influent

A Simple Food Web in a Trickle-Filter System

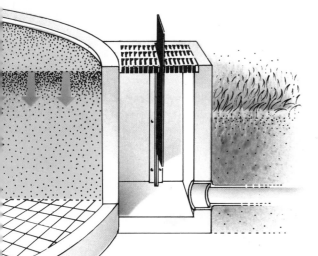

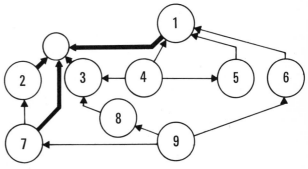

1 Bacteria
2 Amoeba
3 Yeast
4 Rotifer
5 Ciliate
6 Moth fly larva
7 Worm
8 Fly larva
9 Midge larva

Cutaway view of a trickle-filter (above, left) shows raw sewage sprayed onto a hardcore bed. Here, bits of crushed rock (rectangular inset) acquire a sewage coat supporting a food web of tiny organisms (below left). Worms, amoebas, yeast, and bacteria consume sewage and provide food for rotifers, ciliates, and insect larvae. The organisms turn noxious effluent into clear water and other equally harmless waste products.

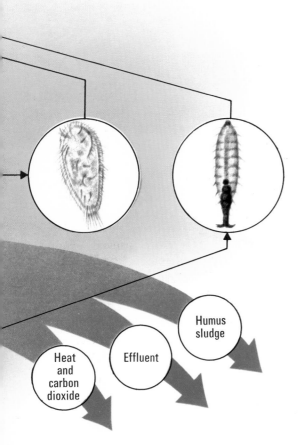

Heat and carbon dioxide

Effluent

Humus sludge

different and highly desirable substances by breaking *down* the substances from which they are derived. So the microorganisms, you might say, are mighty creators as well as destroyers. In no area of modern life does man profit more immediately from this paradox than in the area of sanitation—especially in the purification of water and the disposal of sewage. These are major problems for every modern city, with its insatiable thirst for water and its fantastic ability to produce sewage. For a city of 1 million people, the inflow of water may be as high as 625,000 tons a day, and the output of sewage as much as 500,000 tons a day. Water-treatment plant can remove sewage wastes from water and render it safe to drink; and it is vital that this should be done, for many cities these days are forced to use water that has been discharged through the sewers of other cities farther upstream. Untreated water is dangerous, because of the poisons and some harmful microorganisms in it. So we get rid of them both by means of other, airborne microorganisms.

The first stage in water purification is the removal of solid nonorganic debris—a mechanical operation in which microorganisms play no part. Their role is all-important, however, when it comes to cleansing the water of human and other animal wastes and of virulent microorganisms. This can be done in any one of several ways. In one method of sewage treatment, untreated sewage is run into large tanks, where it remains for up to four weeks. Solid residues sink to the bottom and are periodically removed for burning or burial. Meanwhile, the organic sewage is fermented by bacteria, which produce two gases, carbon dioxide and methane. These bacteria float into the tanks from the atmosphere, but they do not need oxygen to keep alive and multiply. Gradually, all the wastes are converted into carbon dioxide and methane, and in some systems the methane is burned to provide energy for local heating and cooking. The dangerous bacteria sink to the bottom of the tank with the solid debris and are removed and burned.

In another possible method of treatment, the purifying job is done by bacteria that do need oxygen and that accomplish their task in the airy crevices of a six-foot deep bed of crushed rock. Untreated sewage is sprayed onto the top of this enormous filtering device, and as it trickles down through the bed, the wastes stick to the rocks. Naturally microorganisms floating in from the air grow in profusion on these wastes. Additional air is forced through the bed from below. And so the sewage is rendered harmless by what is in effect a miniature food chain. The waste matter is attacked first by slime-forming bacteria, filamentous bacteria, and filamentous fungi; these are in turn eaten by protozoans, which are in their turn eaten by minute animals. By breaking down the wastes to provide themselves with energy, all these hungry creatures leave nothing but harmless carbon dioxide and water. And the water that they leave is thoroughly safe for human consumption.

These and several other sewage-treatment techniques all operate on the same familiar principle of biodegradation. Any sewage-treatment plant, whatever its special methods, is in essence just a collecting place for sewage, where the microbes can most effectively do their work of breaking down organic substances. It is fortunate for us that the microorganisms are so efficient, because a high proportion of our daily water supply depends on them. There are few more instructive experiences than a tour of a sewage plant. If you have seen the state of the water when it comes in for purification and

are handed a glass of it to drink after it emerges from the process, you may momentarily wish you were somewhere else. But you need not hesitate. You can be sure that the busy creatures of the invisible world have done their job.

So if the story of microorganisms sometimes seems overfull of grim details of diseases and destruction, we should remember that the relationship of one organism to another is very finely balanced, and that the damage done to other organisms by the microbes is the exception rather than the rule. In the main, peaceful and mutually beneficial coexistence is the order of the day.

What marks the microorganisms apart from all other life on earth is the strangeness of their tiny existence. This other-worldliness has proved a continual source of fascination to microbiologists ever since Anton van Leeuwenhoek observed and wrote about microscopic life 250 years ago. Although microorganisms impinge upon every aspect of our everyday lives, they seem to us to be a little magical, a little miraculous. The medieval alchemist who longed to change lead into gold would surely have thought himself far along toward his goal if he had been given a private glimpse of the invisible living creatures that can cause and cure disease, turn grape juice into wine, or render poisonous wastes harmless. Even today we marvel at their powers, especially their power to survive. Faced with new situations and new perils, they reproduce so quickly that evolution occurs in a matter of hours rather than thousands of years, and a new breed of microbe almost inevitably manages to overcome or sidestep any dangers.

There is great concern today about our earth. Many fear that it is heading for disaster, whether from the misuse of technology, the destruction of ecological relationships, the population explosion, or the possibility of atomic warfare. No matter what calamity man brings on himself and other visible forms of life, however, the surface of this planet would have to glow red hot, or lie buried under hundreds of feet of ice, before the microorganisms would be cheated out of their unique inheritance.

Crop-spraying from the air controls pests and weeds including microscopic fungi. But while man can exterminate big animals, his invisibly tiny enemies defy total destruction. Microbes— whether hostile, harmless, or beneficial to us—are the most prolific and adaptable of all organisms living on this earth.

Index

Page numbers in *italics* refer to illustrations or captions to illustrations